Lecture Notes in Mathematics

Edited by A. Dold and B. Eckmann

684

T0220047

Elemer E. Rosinger

Distributions and Nonlinear Partial Differential Equations

Springer-Verlag
Berlin Heidelberg New York 1978

Author
Elemer E. Rosinger
Department of Computer Science
Technion City
Haifa/Israel

AMS Subject Classifications (1970): 35 A xx, 35 D xx, 46 F xx

ISBN 3-540-08951-9 Springer-Verlag Berlin Heidelberg New York
ISBN 0-387-08951-9 Springer-Verlag New York Heidelberg Berlin

Printed in Germany

Printing and binding: Beltz Offsetdruck, Hemsbach/Bergstr.
2141/3140-543210

to my wife HERMONA

P R E F A C E

The nonlinear method in the theory of distributions presented in this work is based on embeddings of the distributions in $\mathcal{D}'(R^n)$ into associative and commutative algebras whose elements are classes of sequences of smooth functions on R^n. The embeddings define various distribution multiplications. Positive powers can also be defined for cer tain distributions, as for instance the Dirac δ function.

A framework is in that way obtained for the study of nonlinear partial differential equations with weak or distribution solutions as well as for a whole range of irregular operations on distributions, encountered for instance in quantum mechanics.

In chapter 1, the general method of constructing the algebras containing the distributions and basic properties of these algebras are presented. The way the algebras are constructed can be interpreted as a sequential completion of the space of smooth functions on R^n. In chapter 2, based on an analysis of classes of singularities of piece wise smooth functions on R^n, situated on arbitrary closed subsets of R^n with smooth boundaries, for instance, locally finite families of smooth surfaces, the so called Dirac algebras, which prove to be useful in later applications are introduced.

Chapter 3 presents a first application. A general class of nonlinear partial differential equations, with polynomial nonlinearities is considered. These equations include among others, the nonlinear hyperbolic equations modelling the shock waves as well as well known second order nonlinear wave equations. It is shown that the piece wise smooth weak solutions of the general nonlinear equations considered, satisfy the equations in the usual algebraic sense, with the multiplication and derivatives in the algebras containing the distributions. It follows in particular that the same holds for the piece wise smooth shock wave solutions of nonlinear hyperbolic equations.

A second application is given in chapter 4, where one and three dimensional quantum particle motions in potentials arbitrary positive powers of the Dirac δ function are considered. These potentials which are no more measures, present the strongest local singularities studied in scattering theory. It is proved that the wave function solutions obtained within the algebras containing the distributions, possess the scattering property of being solutions of the potential free equations on either side of the potentials while satisfying special junction relations on the support of the potentials. In chapter 5, relations involving irregular products with Dirac distributions are proved to be valid within the algebras containing the distributions. In particular, several known relations in quantum mechanics, involving irregular products with

Dirac and Heisenberg distributions are valid within the algebras. Chapter 6 presents the peculiar effect coordinate scaling has on Dirac distribution derivatives. That effect is a consequence of the condition of strong local presence the representations of the Dirac distribution satisfy in certain algebras. In chapter 7, local properties in the algebras are presented with the help of the notion of support, the local character of the product being one of the important results. Chapter 8 approaches the problem of vanishing and local vanishing of the sequences of smooth functions which generate the ideals used in the quotient construction giving the algebras containing the distributions. That problem proves to be closely connected with the necessary structure of the distribution multiplications. The method of sequential completion used in the construction of the algebras containing the distributions establishes a connection between the nonlinear theory of distributions presented in this work and the theory of algebras of continuous functions.

The present work resulted from an interest in the subject over the last few years and it was accomplished while the author was a member of the Applied Mathematics Group within the Department of Computer Science at Haifa Technion. In this respect, the author is particularly glad to express his special gratitude to Prof. A. Paz, the head of the department, for the kind support and understanding offered during the last years.

Many thanks go to the colleagues at Technion, M. Israeli and L. Shulman, for valuable reference indications, respectively for suggesting the scattering problem in potentials positive powers of the Dirac δ function, solved in chapter 4.

The author is indebted to Prof. B. Fuchssteiner from Paderborn, for his suggestions in contacting persons with the same research interest.

Lately, the author has learnt about a series of extensive papers of K. Keller, from the Institute for Theoretical Physics at Aachen, presenting a rather complementary approach to the problem of irregular operations with distributions. The author is very glad to thank him for the kind and thorough exchange of views.

A special gratitude and acknowledgement is expressed by the author to R.C. King from Southampton University, for his generosity in promptly offering the result on generalized Vandermonde determinants which corrects an earlier conjecture of the author and upon which the chapters 5 and 6 are based.

All the highly careful and demanding work of editing the manuscript was done by my wife Hermona, who inspite and on the account of her other much more interesting and elevated usual occupations found it necessary to support an effort in regularizing the irregulars ..., in multiplying the distributions ...

By the way of multiplication: Prof. A. Ben-Israel, a former colleague, noticing the series of preprints, papers, etc. resulted from the author's interest in the subject and seemingly inspired by one of the basic commandments in the Bible, once quipped: "Be fruitful and multiply ... distributions ..."

E. E. R.

Haifa, December 1977

CONTENT

N O T E

The Reader interested mainly in NONLINEAR PARTIAL DIFFERENTIAL EQUATIONS, may at a first lecture concentrate on the following sections:

"Never forget

the beaches of ASHQELON ... "

Chapter 1

ASSOCIATIVE, COMMUTATIVE ALGEBRAS CONTAINING THE DISTRIBUTIONS

§1. NONLINEAR PROBLEMS

The theory of distributions has proved to be essential in the study of linear partial differential equations. The general results concerning the existence of elementary solutions, [103], [34], P-convexity as the necessary and sufficient condition for the existence of smooth solutions, [103], the algebraic characterization of hypoellipticity, [64], etc., are several of the achievements due to the distributional approach, [154], [63], [64], [153], [156], [33], [114].

In the case of nonlinear partial differential equations certain facts have pointed out the useful role a nonlinear theory of distributions could play. For instance, the appearance of shock discontinuities in the solutions of nonlinear hyperbolic partial differential equations, even in the case of analytic initial data, [62], [89], [113], [50] [70], [51], [24], [25], [26], [31], [32], [52], [58], [71], [79], [84], [90], [91], [133], [149], [163], indicates that in the nonlinear case problems arise starting with a rigorous and general definition of the notion of solution. Important cases of nonlinear wave equations, [5], [9], [10], [11], [121], prove to possess distribution solutions of physical interest, provided that 'irregular' operations, e.g. products, with distributions are defined. Using suitable procedures, distribution solutions can be associated to various nonlinear differential or partial differential equations, [1], [2] [30], [42], [43], [45], [80], [92], [94], [117], [118], [119], [120], [122], [138], [146], [147], [159], [160], [161].

In quantum mechanics, procedures of regularizing divergent expressions containing 'irregular' operations with distributions, such as products, powers, convolutions, etc., have been in use, [6], [7], [12], [13], [15], [19], [20], [21], [29], [54], [55], [56] [57], [60], [76], [77], [78], [112], [143], [151], suggesting the utility of enriching in a systematic way the vector space structure of the distributions.

A natural way to start a nonlinear theory of distributions is by supplementing the vector space structure of $\mathcal{D}'(R^n)$ with a suitable distribution multiplication.

Within this work, a nonlinear method in the theory of distributions is presented, based on an associative and commutative multiplication defined for the distributions in $\mathcal{D}'(R^n)$, [125-131]. That multiplication offers the possibility of defining arbitrary positive powers for certain distributions, e.g. the Dirac δ function, [130], [151].

The definition of the multiplication rests upon an analysis of classes of singularities of piece wise smooth functions on R^n, situated on arbitrary closed subsets of R^n with smooth boundaries, for instance locally finite families of smooth surfaces in R^n (chap. 2, § 3).

Several applications are presented.
First, in chapter 3, it is shown that the piece wise smooth weak solutions of a general class of nonlinear partial differential equations satisfy those equations in the usual algebraic sense, with the multiplication and derivatives in the algebras containing the distributions. As a particular case, it results that the piece wise smooth nonlinear shock wave solutions of the equation, [90], [71], [133], [52], [32] [131]:

$$u_t(x,t) + a(u(x,t)) \cdot u_x(x,t) = 0 \ , \ x \in R^1 \ , \ t > 0 \ ,$$
$$u(x,0) = u_0(x) \ , \ x \in R^1 \ ,$$

where a is an arbitrary polynomial in u, satisfy that equation in the usual algebraic sense.

Second, in chapter 4, quantum particle motions in potentials arbitrary positive powers of the Dirac δ distribution are considered. These potentials present the strongest local singularities studied in recent literature on scattering, [27], [3], [28], [115], [116], [140]. The one dimensional motion has the wave function ψ given by

$$\psi''(x) + (k-\alpha(\delta(x))^m)\psi(x) = 0 \ , \ x \in R^1 \ , \ k,\alpha \in R^1 \ , \ m \in (0,\infty]$$

while the three dimensional motion assumed spherically symmetric and with zero angular momentum has the radial wave function R given by

$$(r^2R'(r))' + r^2(k-\alpha(\delta(r-a))^m)R(r) = 0 \ , \ r \in (0,\infty) \ , \ k,\alpha \in R^1 \ , \ a,m \in (0,\infty) \ .$$

The wave function solutions obtained possess a usual scattering property, namely they consist of pairs ψ_-,ψ_+ of usual C^∞ solutions of the potential free equations, each valid on the respective side of the potential while satisfying special junction relations on the support of the potentials.

Third, it is shown in chap. 5, §5, that the following well known relations in quantum mechanics, [108], involving the square of the Dirac δ and Heisenberg δ_+,δ_- distributions and other irregular products hold:

$$(\delta)^2 - (1/x)^2/\pi^2 = -(1/x^2)/\pi^2$$

$$\delta \cdot (1/x) = -D\delta/2$$

$$(\delta_+)^2 = -D\delta/4\pi i - (1/x^2)/4\pi^2$$

$$(\delta_-)^2 = D\delta/4\pi i - (1/x^2)/4\pi^2$$

where $\delta_+ = (\delta+(1/x)/\pi i)/2$, $\delta_- = (\delta-(1/x)/\pi i)/2$.

§2. MOTIVATION OF THE APPROACH

The distribution multiplication, defined for any given pair of distributions in $\mathcal{D}'(R^n)$, could either lead again to a distribution or to a more general entity. Taking into account H. Lewy's simple example, [93] (see also [64], [155], [48]), of a first order linear partial differential operator with three independent variables and coefficients polynomials of degree at most one with no distribution solutions, the choice of a distribution multiplication which could in the case of particularly irregular factors lead outside of the distributions, seems worthwhile considering. Such an extension beyond the distributions would mean an increase in the 'reservoir' of both data and possible solutions of nonlinear partial differential operators, not unlike it happened with the introduction of distributions in the study of linear partial differential operators, [154].

One can obtain a distribution multiplication in line with the above remarks by embedding $\mathcal{D}'(R^n)$ into an algebra A *). It would be desirable for a usual Calculus if the algebra A were associative, commutative, with the function $\psi(x) = 1$, $\forall x \in R^n$, its unit element and possessing derivative operators satisfying Leibnitz type rules for the product derivatives. Certain supplementary properties of the embedding $\mathcal{D}'(R^n) \subset A$ concerning multiplication, derivative, etc. could also be envisioned.

There is a particularly convenient classical way to obtain such an algebra A , namely, as a sequential completion of $\mathcal{D}'(R^n)$ or eventually, of a subspace F in $\mathcal{D}'(R^n)$. The sequential completion, suggested by Cauchy and Bolzano, [158], was employed rigorously by Cantor, [22], in the construction of R^1 . Within the theory of distributions the sequential completion was first employed by J. Mikusinski, [105] (see also [110]) in order to construct the distributions in $\mathcal{D}'(R^1)$ from the set of locally integrable functions on R^1 , however without aiming at defining a distribution multiplication.

---

*) All the algebras in the sequel are considered over the field C^1 of the complex numbers.

Later, in [106], the problem of a whole range of 'irregular' operations - among them, multiplication - was formulated within the framework of the sequential completion.

The method of the sequential completion possesses two important advantages. First, there exist various subspaces F in $\mathcal{D}'(R^n)$ which are in a natural way associative, commutative algebras, with the unit element the function $\psi(x) = 1$, $\forall\, x \in R^n$. Starting with such a subspace F , it is easy to construct a sequential completion A which will also be an associative, commutative algebra with unit element. Indeed, the procedure is - from purely algebraic point of view - the following one. Denote $\mathcal{W} = N \to F$, that is, the set of all sequences with elements in F . With the term by term operations on sequences, $\mathcal{W}$ is an associative, commutative algebra with unit element. Choosing a suitable subalgebra $\mathcal{A}$ in $\mathcal{W}$ and an ideal in $\mathcal{A}$, one obtains $A = \mathcal{A}/\mathcal{I}$.

Second, the sequential completion A results in a constructive way. Further, a simple characterization of the elements in A is obtained. Indeed, these elements will be classes of sequences of 'regular' functions in R^n (in this work, $F = \mathcal{C}^\infty(R^n)$ will be considered) much in the spirit of various 'weak solutions' used in the study of partial differential equations.

Within the more general framework of Calculus, the distributional approach - essentially a sequential completion of a function space, [105],[110],[4] - can be viewed as a stage in a succession of attempts to define the notion of function. Euler's idea of function, as an analytic one was extended by Dirichlet's definition accepting any univalent correspondence from numbers to numbers. That extension although significant - encompassing even nonmeasurable functions, provided the Axiom of Choice is assumed, [49] - failed to include certain rather simple important cases, as for instance, the Dirac δ function and its derivatives.

It is worthwhile mentioning that the distributional approach can be paralleled by certain approaches in Nonstandard Analysis. In [134], a nonstandard model of R^1 obtained by a sequential completion of the rational numbers was presented. In that nonstandard R^1 , the Dirac δ function becomes a usual univalent correspondence from numbers (nonstandard) to numbers (nonstandard).

The notion of the germ of a function at a point which can be regarded as a generalization of the notion of function, since it represents more than the value of the function at the point but less than the function on any given neighbourhood of the point, is related both to the distributional approach and Nonstandard Analysis, [109],[97].

The variety of interrelated approaches suggests that the notion of function in Calculus is still 'in the making'. The particular success of the distributional approach in the theory of linear partial differential equations (especially the constant coefficient case, otherwise see [93]) is in a good deal traceable to the strong results and methods in linear functional analysis and functions of several complex variables. In this respect, the distributional approach of nonlinear problems, such as nonlinear partial differential equations, can be seen as requiring a return to more basic and general methods, as for instance, the sequential completion of convenient function spaces, which finds a natural framework in the theory of Algebras of Continuous Functions (see chap. 8).

The sequential completion is a common method for both standard and nonstandard methods in Calculus and its theoretical importance is supplemented by the fact that it synthetizes basic approximation methods used in applications, such as the method of 'weak solutions'. The nonlinear method in the theory of distributions presented in this work is based on the embedding of $\mathcal{D}'(R^n)$ into associative and commutative algebras with unit element, constructed by particular sequential completions of $\mathcal{C}^\infty(R^n)$, resulting from an analysis of classes of singularities of piece wise smooth functions on R^n, situated on arbitrary closed subsets of R^n with smooth boundaries, for instance locally finite families of smooth surfaces in R^n (see chap. 2, §3).

§3. DISTRIBUTION MULTIPLICATION

The problem of distribution multiplication appeared early in the theory of distributions, [135], [81-83], and generated a literature, [9], [11-21], [35-41], [46], [53-57], [61], [66-69], [72-74], [76-78], [85], [106-108], [112], [125-132], [134], [136] [137], [148], [151], [162]. L. Schwartz's paper [135], presented a first account of the difficulties. Namely, it was shown impossible to embed $\mathcal{D}'(R^1)$ into an associative algebra A under the following conditions:

a) the function $\psi(x) = 1$, $\forall\, x \in R^1$, is the unit element of the algebra A ;

b) the multiplication in A of any two of the functions

$$1, x, x(\ell n|x|-1) \in \mathcal{C}^0(R^1)$$

is identical with the usual multiplication in $\mathcal{C}^0(R^1)$;

c) there exists a linear mapping (generalized derivative operator) $D : A \to A$, such that:

c.1) D satisfies on A the Leibnitz rule of product derivative:

$$D(a\cdot b) = (Da)\cdot b + a\cdot(Db)\ ,\quad \forall\, a,b \in A\ ;$$

c.2) D applied to the functions

$$1, x, x^2(\ell n|x|-1) \in C^1(R^1)$$

is identical with the usual derivative in $C^1(R^1)$;

d) there exists $\delta \in A$, $\delta \neq 0$ (corresponding to the Dirac function) such that $x \cdot \delta = 0$.

The above negative result was occasionally interpreted as amounting to the impossibility of a useful distribution multiplication. That could have implied that the distributional approach was not suitable for a systematic study of nonlinear problems. However, due to applicative interest (see §1) various distribution multiplications satisfying on the one side weakened forms of the conditions in [135] but, now and then also rather strong and interesting other conditions not considered in [135], have been suggested and used as seen in the above mentioned literature. In this respect, the challenging question keeping up the interest in distribution multiplication has been the following one: which sets of strong and interesting properties can be realized in a distribution multiplication?

There has been as well an other source of possible concern, namely, the rather permanent feature of the distribution multiplications suggested, that the product of two distributions with significant singularities can contain arbitrary parameters. However, a careful study of various applications shows that the parameters can be in a way or the other connected with characteristics of the particular nonlinear problems considered. The complication brought in by the lack of a unique, so called 'canonical' product, and the 'branching' the multiplication shows above a certain level of singularities can be seen as a rather necessary phenomenon accompanying operations with singularities.

The study of the literature on distribution multiplications points out two main approaches. One of them tries to define for as many distributions as possible, products which are again distributions, [9], [11-21], [35-41], [46], [53-57], [61], [66-69], [72-74], [78], [106-108], [112], [132], [136], [137], [148], [162]. That approach can be viewed as an attempt to construct maximal 'subalgebras' in $\mathcal{D}'(R^n)$, using various regularization procedures applied to certain linear functionals associated to products of distributions. Sometimes, [9], [11], [78], the regularization procedures are required to satisfy certain axioms considered to be natural. A general characteristic of the approach is a trade-off between the primary aim of keeping the multiplication within the distributions and the resulting algebraic and topological properties of the multiplication which prove to be weaker than the ones within the usual algebras of functions or operators. The question arising connected with that approach is whether the advantage of keeping the product within the distributions compensates for the resulting restrictions on operations as well as for the lack of properties customary in a good Calculus.

The other approach, a rather complementary one, aimes first to obtain a rich algebraic structure with suitable derivative operators, enabling a Calculus with minimal restrictions, [81-83], [76], [77], [85], [125-131], [134], [151]. That approach can be seen as an attempt to construct embeddings of $\mathcal{D}'(R^n)$ into algebras.

The present work belongs to the latter approach.

A more fair comment would perhaps say that within the first approach, one knows what he computes with, even if not always how to do it, while within the second approach, one easily knows how to compute, even if not always what the result is. However, the second approach seems to be more in line with the initial spirit of the Theory of Distributions, aiming at lifting restrictions, simplifying rules and extending the ranges of operations in Calculus, even if done by adjoining unusual entities.

§4. ALGEBRAS OF SEQUENCES OF SMOOTH FUNCTIONS

The set

(1) $$\mathcal{W} = N \to \mathcal{C}^\infty(R^n)$$

of all the sequences of complex valued smooth functions on R^n will give in the sequel the general framework. If $s \in \mathcal{W}$, $\nu \in N$, $x \in R^n$, then $s(\nu) \in \mathcal{C}^\infty(R^n)$ and $s(\nu)(x) \in C^1$.

For $\psi \in \mathcal{C}^\infty(R^n)$ denote by $u(\psi)$ the constant sequence with the terms ψ , then $u(\psi) \in \mathcal{W}$ and $u(\psi)(\nu) = \psi$, $\forall \nu \in N$.

With the term by term addition and multiplication of sequences, $\mathcal{W}$ is an associative, comutative algebra with the unit element $u(1)$; the null subspace of $\mathcal{W}$ is $\mathcal{O} = \{u(0)\}$.

Denote by $\mathcal{S}_0$ the set of all sequences $s \in \mathcal{W}$, weakly convergent in $\mathcal{D}'(R^n)$ and by $\mathcal{V}_0$ the kernel of the linear surjection:

(2) $$\mathcal{S}_0 \ni s \longrightarrow \langle s, \cdot \rangle \in \mathcal{D}'(R^n)$$

where $\langle s,\psi\rangle = \lim_{\nu\to\infty} \int_{R^n} s(\nu)(x)\psi(x)dx$, $\forall \psi \in \mathcal{D}(R^n)$.

Then

(3) $$\mathcal{S}_0/\mathcal{V}_0 \ni (s+\mathcal{V}_0) \xrightarrow{\ \omega\ } \langle s,\cdot\rangle \in \mathcal{D}'(R^n)$$

is a vector space isomorphism.

Since W is an algebra, one can ask whether it is possible to define a product of any two distributions $\langle s,\cdot\rangle$, $\langle t,\cdot\rangle \in \mathcal{D}'(R^n)$ by the product of the classes of sequences $s + V_0$ and $t + V_0$.

A simple way to do it would be by constructing diagrams of inclusions:

$$
(4) \qquad
\begin{array}{ccccc}
I & \longrightarrow & A & \longrightarrow & W \\
\uparrow & & \uparrow & & \\
V_0 & \longrightarrow & S_0 & &
\end{array}
$$

with A subalgebra in W and I ideal in A , satisfying

(4.1) $\qquad I \cap S_0 = V_0$

which would generate the following linear embedding of $\mathcal{D}'(R^n)$ into an associative and commutative algebra with unit element:

$$
\begin{array}{ccccc}
\mathcal{D}'(R^n) & & S_0/V_0 & & A/I \\
\cup\!\!\!\!\mid & & \cup\!\!\!\!\mid & & \cup\!\!\!\!\mid \\
\langle s,\cdot\rangle & \underset{\text{isom}}{\longleftarrow} & s+V_0 & \underset{\text{lin,inj}}{\longrightarrow} & s+I
\end{array}
$$

However, diagrams of type (4) cannot be constructed, since

(5) $\qquad (V_0 \cdot V_0) \cap S_0 \not\subset V_0$

Indeed, if $n = 1$, take $v(\nu)(x) = \cos(\nu+1)x$, $\forall\, \nu \in N$, $x \in R^1$, then $v \in V_0$, $v^2 \in S_0$ and $v^2 \notin V_0$, since $\langle v^2,\cdot\rangle = 1/2$.

An other way could be given by diagrams of inclusions:

$$
(6) \qquad
\begin{array}{ccccc}
I & \longrightarrow & A & \longrightarrow & W \\
\downarrow & & \downarrow & & \\
V_0 & \longrightarrow & S_0 & &
\end{array}
$$

with A subalgebra in W and I ideal in A , satisfying

(6.1) $\qquad V_0 \cap A = I$

(6.2) $\qquad V_0 + A = S_0$

which would generate the following linear injection of an associative and commutative algebra onto $\mathcal{D}'(R^n)$:

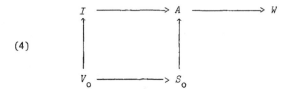

$$\begin{array}{ccccc} D'(R^n) & & S_o/V_o & & A/I \\ \rotatebox{90}{$\in$} & & \rotatebox{90}{$\in$} & & \rotatebox{90}{$\in$} \\ \langle s,\cdot\rangle & \xleftarrow[\text{isom}]{\hspace{3em}} & s+V_o & \xleftarrow[\text{lin,inj,sur}]{\hspace{3em}} & s+I \end{array}$$

Here the problem arises connected with (6.2). Indeed, it is not possible to construct diagrams of type (6) with A containing some of the frequently used 'δ sequences', [4], [35-41], [53], [68], [69], [105-110], [136], [137], [162], as results from (see the proof in §12):

Lemma 1

Suppose given $s \in W$ such that supp $s(\nu)$ shrinks to $0 \in R^n$, when $\nu \to \infty$. Then

1) In case s is a sequence of nonnegative functions, the following two properties are equivalent

   1.1) $s \in S_o$ and $\langle s,\cdot\rangle = \delta$ (the Dirac distribution)

   1.2) $\lim\limits_{\nu\to\infty} \int_{R^n} s(\nu)(x)dx = 1$

2) If $s \in S_o$ and $\langle s,\cdot\rangle = \delta$, then $s^2 \notin S_o$

The importance of the above type of 'δ sequences' is due to the smooth approximations they generate for functions $f \in L^1_{loc}(R^n)$ through the convolutions $f_\nu = f * s(\nu)$, $\nu \in N$.

In [125-131] it was shown (see also Theorem 1, §7) that the following, slightly more complicated diagrams of inclusions can be constructed:

$$(7) \qquad \begin{array}{ccccc} I & \longrightarrow & A & \longrightarrow & W \\ \uparrow & & \uparrow & & \\ V & \longrightarrow & S \ni u(1) & & \\ \downarrow & & \downarrow & & \\ V_o & \longrightarrow & S_o & & \end{array}$$

with A subalgebra in W, I ideal in A and V, S vector subspaces in S_o, satisfying the conditions:

(7.1) $\quad I \cap S = V$

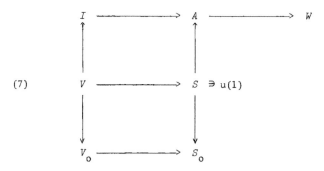

(7.2) $\quad V_0 \cap S = V$

(7.3) $\quad V_0 + S = S_0$

and thus generating the following linear embedding of $\mathcal{D}'(\mathrm{R}^n)$ into an associative and commutative algebra with unit element:

(8)
$$\begin{array}{ccccccc} \mathcal{D}'(\mathrm{R}^n) & & S_0/V_0 & & S/V & & A/I \\ \rotatebox{90}{$\in$} & & \rotatebox{90}{$\in$} & & \rotatebox{90}{$\in$} & & \rotatebox{90}{$\in$} \\ \langle s,\cdot\rangle & \xleftarrow[\text{isom}]{\hspace{2cm}} & s+V_0 & \xleftarrow[\text{isom}]{\hspace{2cm}} & s+V & \xrightarrow[\text{lin,inj}]{\hspace{2cm}} & s+I \end{array}$$

The intermediate quotient space S/V has the role of a regularization of the representation of the distributions in $\mathcal{D}'(\mathrm{R}^n)$ by classes of sequences of weakly convergent smooth functions, given in (3).

§5. SIMPLER DIAGRAMS OF INCLUSIONS

In constructing diagrams of inclusions of type (7), the main problem proves to be the choice of the regularizing quotient space S/V . One can think of reducing that problem to the choice of S only, since V can be obtained from (7.2). However, it will be more convenient to consider (see (20.2) and Remark D in §7):

(9) $\quad S = V \oplus S'$

with S' vector subspace in S_0 and to replace the problem of the choice of S by the one of the choice of the pair (V,S') . In that case, the conditions (7.2) and (7.3) become

(10) $\quad S_0 = V_0 \oplus S'$

Now, the difficult task remains to fulfil (7.1). Obviously, if any ideal I satisfies (7.1) then the smallest ideal containing V will still satisfy that relation. In this respect, taking in (7) the smallest possible I will be convenient when one constructs the algebras containing the distributions. The smallest I can easily be obtained since $u(1) \in S$ in (9). Indeed, denoting

(11) $\quad \mathrm{I}(V,A)$ = the ideal in A generated by V

one obtains the smallest ideal in A which contains V . Moreover, $\mathrm{I}(V,A)$ being the vector subspace generated by $V \cdot A$ in A , one obtains a particularly useful insight into the structure of the algebra $A/\mathrm{I}(V,A)$ containing the distributions.

Therefore, the diagrams (7) will be considered under the particular form

$$
(12) \qquad
\begin{array}{ccccc}
I(V,A) & \longrightarrow & A & \longrightarrow & W \\
\uparrow & & \uparrow & & \\
V & \longrightarrow & V \oplus S' \ni u(1) & & \\
\downarrow & & \downarrow & & \\
V_o & \longrightarrow & S_o & &
\end{array}
$$

with

(12.1) $\quad V_0 \oplus S' = S_0$

(12.2) $\quad I(V,A) \cap (V \oplus S') = V$

It will be useful to notice that (12.2) can be written under the equivalent simpler form:

(12.3) $\quad I(V,A) \cap S' = 0$

In the case of the diagrams (12), the embeddings (8) will obtain the particular form

$$
(13) \qquad
\begin{array}{ccccccc}
D'(R^n) & & S_0/V_0 & & V \oplus S'/V & & A/I(V,A) \\
\ni & & \ni & & \ni & & \ni \\
\langle s,\cdot\rangle & \underset{\text{isom}}{\longleftarrow} & s+V_0 & \underset{\text{isom}}{\longleftarrow} & s+V & \underset{\text{lin,inj}}{\longrightarrow} & s+I(V,A)
\end{array}
$$

Besides the problem of choosing (V,S'), it apparently remains the problem of choosing A. However, that latter problem will be solved in §7, in an easy way. Therefore the problem of embedding the distributions in $D'(R^n)$ into algebras will be reduced to the problem of constructing suitable regularizations (V,S').

§6. ADMISSIBLE PROPERTIES

Several properties of the algebras $A/I(V,A)$, such as the existence of derivative operators on the algebras, the existence of positive powers for certain elements of the algebras, etc. will depend on corresponding properties of the algebras A. A uniform approach of these properties can be obtained with the help of the following definition. A property P, valid for certain subsets H in W is called <u>admissible</u>, only if

(14.1) $\quad W$ has the property P,

(14.2) $\quad$ an intersection of subsets in W, each having the property P, will also have that property.

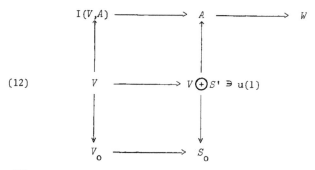

Suppose, P and Q are admissible properties, then P is called <u>stronger</u> than Q and denoted $P \geq Q$, only if each subset H in W satisfying P will also satisfy Q . Obviously, if $P_1,\dots,P_m$ are admissible properties, then their conjunction $P = (P_1 \text{ and } \dots \text{ and } P_m)$ is also an admissible property and $P = \max\{P_1,\dots,P_m\}$ with the above partial order $\geq$.

Denote by $\bar{P}$ the property of subsets H in W that $H = W$, then obviously $\bar{P}$ is the <u>strongest</u> admissible property.

Three of the admissible properties of subsets $H \subset W$ encountered in the sequel are defined now.

(15) H is <u>derivative invariant</u>, only if

(15.1) $D^p H \subset H \ , \quad \forall \ p \in N^n$

where $D^p : W \to W$ is the term by term p-order derivative of the sequences in W , that is, $(D^p s)(\nu)(x) = (D^p s(\nu))(x) \ , \quad \forall \ s \in W \ , \quad \nu \in N \ , \quad x \in R^n$.

(16) H is <u>positive power invatiant</u>, only if

(16.1) $\forall \ s \in H$:

$$\left(\begin{array}{l} *) \quad s(\nu)(x) \geq 0 \ , \quad \forall \ \nu \in N \ , \quad x \in R^n \\ **) \quad s^\alpha \in W \ , \quad \forall \ \alpha \in (0,\infty) \end{array} \right) \Rightarrow (s^\alpha \in H \ , \quad \forall \ \alpha \in (0,\infty))$$

where $s^\alpha(\nu)(x) = (s(\nu)(x))^\alpha \ , \quad \forall \ \nu \in N \ , \quad x \in R^n$.

(17) H is <u>sectional invariant</u>, only if

(17.1) $\forall \ t \in W$:

$$\left(\begin{array}{l} \exists \ s \in H \ , \quad \mu \in N : \\ \forall \ \nu \in N \ , \quad \nu \geq \mu : \\ \qquad t(\nu) = s(\nu) \end{array} \right) \Rightarrow t \in H$$

A related, in a way stronger admissible property is defined in:

(17') H is <u>subsequence invariant</u>, only if any subsequence t of a sequence $s \in H$ is also belonging to H .

§7. REGULARIZATIONS AND ALGEBRAS CONTAINING THE DISTRIBUTIONS

The aim of this section is to define general classes of regularizations and to construct the corresponding algebras containing $\mathcal{D}'(R^n)$.

The construction of the algebras is carried out assuming that a certain admissible property P was specified in advance.

Suppose now, given any admissible porperty Q, such that $Q \le P$.

For (V,S') given in §5, the choice of the subalgebra A needed in the diagram (12) in order to obtain the embedding (13), will be

(18) $A^Q(V,S')$ = the smallest subalgebra in W with the property Q and containing $V \oplus S'$.

The notation (see(11))

(19) $$I^Q(V,S') = I(V,A^Q(V,S'))$$

will be useful.

In Theorem 1, below, it is proved that one can reduce the construction of diagrams (12) to the choice of pairs given in the following:

Definition

A pair (V,S') of vector subspaces in V_0 respectively S_0 is called a P-regularization, only if

(20.1) $$S_0 = V_0 \oplus S'$$

(20.2) $$U \subset V(p) \oplus S' \; , \quad \forall\, p \in \bar{N}^n \quad ^{*)}$$

(20.3) $$I^P(V,S') \cap S' = 0$$

where

(21) $$U = \{u(\psi) \mid \psi \in C^\infty(R^n)\}$$

is the set of constant sequences of smooth functions, and

(22) $$V(p) = \{v \in V \mid \forall\, r \in N^n \, , \; r \le p : D^r v \in V\} \, , \quad \forall\, p \in \bar{N}^n \, .$$

Denote by $R(P)$ the set of all P-regularizations (V,S').

If Q is an admissible property and $Q \le P$ then $R(P) \subset R(Q)$, due to (20.3), (19) and (18). Therefore, $R(\bar{P}) \subset R(P)$ for any admissible property P, since $\bar{P}$ is the strongest admissible property (see §6).

A pair $(V,S') \in R(\bar{P})$ will be called regularization. Thus, a (V,S') is regularization, only if it is a P-regularization, for any admissible property P.

---

*) $\bar{N} = N \cup \{\infty\}$

Remark 1

1) The above condition (20.1) is identical with (12.1) while (20.3) is equivalent with (12.3), assuming that one takes in (12), $A = A^{P}(V,S')$. The condition (20.2) is a stronger version of the relation $u(1) \in V \oplus S'$ in (12) and it is needed in order to secure the fact that the multiplication in the algebras containing D' induces on C^{∞} the usual multiplication of functions. The presence of $V(p)$, with $p \in \bar{N}^n$, in (2o.2) is connected with the family of algebras mentioned in Remark D, below, used in order to define proper derivative operators.

2) If (V,S') is a P-regularization then $V \subsetneqq V_0$. Indeed, assume $V = V_0$, then (20.1) and (20.3) result in $I^{P}(V,S') \cap S_0 \subset V_0$. But (19) implies $V_0 \subset I^{P}(V,S')$, therefore (5) is contradicted.

Remark D

It is important to point out the necessary connection between the way the derivative operators are defined on the algebras and the validity of certain basic relations involving important distributions. Indeed, as seen in §11, assuming:

a) the existence of an algebra $A \supset D'(R^1)$ possessing a derivative operator $D : A \to A$, and

b) the validity of the relation $x \cdot \delta = 0$,

one necessarily obtains $\delta^2 = 0$, a relation not always in line with the possible interpretations of δ^2 (see chap. 4 and [11], [21], [108], [151]) (the relation $x \cdot \delta = 0$ is important in that it gives an upper bound of the order of singularity the Dirac δ function exhibits at $0 \in R^1$) .

A way out is to embed $D'(R^n)$ into a <u>family of algebras</u> A_p , with $p \in \bar{N}^n$, possessing derivative operators (see Theorem 3, §8):

$$D^q : A_{p+q} \longrightarrow A_p , \quad \forall\ q \in N^n , \quad p \in \bar{N}^n$$

From here the presence of the vector subspaces $V(p)$, with $p \in \bar{N}^n$, in the condition (20.2). However, that method can still lead to the situation in a) above, provided that $D^q V \subset V$, $\forall\ q \in N^n$, in which case the algebras A_p with $p \in \bar{N}^n$ will be identical.

And now, the basic result in the present chapter.

Theorem 1

$R(P)$ is not void.

Suppose given $(V,S') \in R(P)$ and an admissible property Q , such that $Q \leq P$.

Then, for each $p \in \bar{N}^n$, the diagram of inclusions holds

$$
(23) \qquad
\begin{array}{ccccc}
I^Q(V(p),S') & \longrightarrow & A^Q(V(p),S') & \longrightarrow & W \\
\uparrow & & \uparrow & & \\
V(p) & \longrightarrow & V(p) \oplus S' & \longleftarrow & U \\
\downarrow & & \downarrow & & \\
V_0 & \longrightarrow & S_0 & &
\end{array}
$$

and

(23.1) $\qquad I^Q(V(p),S') \cap (V(p) \oplus S') = V(p)$,

or, equivalently

(23.2) $\qquad I^Q(V(p),S') \cap S' = 0$

Proof

$R(P)$ is not void due to Theorem 4 in chap. 2, §6.

The inclusions in (23) result easily and we shall only prove (23.1). Obviously, $V(p) \subset V$, hence $A^Q(V(p),S') \subset A^Q(V,S')$. Noticing that due to $Q \leq P$, the inclusion $A^Q(V,S') \subset A^P(V,S')$ holds, one obtains $I^Q(V(p),S') \subset I^P(V,S')$. Therefore, $I^Q(V(p),S') \cap (V(p) \oplus S') \subset I^P(V,S') \cap (V \oplus S') \subset V_0$, the last inclusion resulting from (20.3). Now, obviously $I^Q(V(p),S') \cap (V(p) \oplus S') \subset V_0 \cap (V(p) \oplus S') = V(p)$ and the inclusion $\subset$ in (23.1) is proved. The converse inclusion resulting from (23), the proof of Theorem 1 is completed $\nabla\nabla\nabla$

And now, the definition of the family of associative and commutative algebras with unit element, each containing the distributions in $D'(R^n)$, family associated to any given P-regularization.

Suppose $(V,S') \in R(P)$ and Q is an admissible property, with $Q \leq P$. Denote then

(24) $\qquad A^Q(V,S',p) = A^Q(V(p),S')/I^Q(V(p),S')$, with $p \in \bar{N}^n$.

The algebras $A^Q(V,S',p)$ will be called derivative algebras, positive power algebras, sectional algebras or subsequence algebras, only if Q is respectively stronger than the admissible properties (15), (16), (17) or (17').

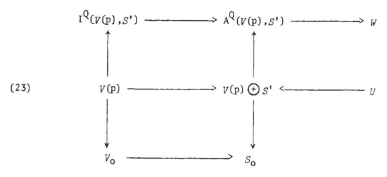

§8. PROPERTIES OF THE FAMILIES OF ALGEBRAS CONTAINING $\mathcal{D}'(\mathbb{R}^n)$

The next three theorems present the main properties of the embeddings $\mathcal{D}'(\mathbb{R}^n) \subset A^Q(\mathcal{V},\mathcal{S}',p)$.

Theorem 2

Suppose given $(\mathcal{V},\mathcal{S}') \in \mathcal{R}(P)$ and an admissible property Q , such that $Q \leq P$. Then

1) $A^Q(\mathcal{V},\mathcal{S}',p)$ is an associative, commutative algebra with the unit element $u(1) + I^Q(\mathcal{V}(p),\mathcal{S}')$, for each $p \in \bar{\mathbb{N}}^n$.

2) The following linear applications exist for each $p \in \bar{\mathbb{N}}^n$:

$$\mathcal{S}_0/\mathcal{V}_0 \xleftarrow[\text{bij}]{\alpha_p} \mathcal{V}(p) \oplus \mathcal{S}'/\mathcal{V}(p) \xrightarrow[\text{inj}]{\beta_p} A^Q(\mathcal{V},\mathcal{S}',p)$$

with $\alpha_p(s+\mathcal{V}(p)) = s + \mathcal{V}_0$

$\beta_p(s+\mathcal{V}(p)) = s + I^Q(\mathcal{V}(p),\mathcal{S}')$

3) For each $p \in \bar{\mathbb{N}}^n$, the linear injective application (embedding) exists:

(25)
$$\mathcal{D}'(\mathbb{R}^n) \xrightarrow{\varepsilon_p} A^Q(\mathcal{V},\mathcal{S}',p)$$

with $\varepsilon_p = \beta_p \circ \alpha_p^{-1} \circ \omega^{-1}$ (see (3))

4) For each $p \in \bar{\mathbb{N}}^n$, the multiplication in $A^Q(\mathcal{V},\mathcal{S}',p)$ induces on $C^\infty(\mathbb{R}^n)$ the usual multiplication of functions.

5) For each $p,q,r \in \bar{\mathbb{N}}^n$, $p \leq q \leq r$, the diagram of algebra homomorphisms is commutative:

$$\begin{array}{ccccc} & & \gamma_{r,p} & & \\ A^Q(\mathcal{V},\mathcal{S}',r) & \xrightarrow[\gamma_{r,q}]{} & A^Q(\mathcal{V},\mathcal{S}',q) & \xrightarrow[\gamma_{q,p}]{} & A^Q(\mathcal{V},\mathcal{S}',p) \end{array}$$

with $\gamma_{q,p}(s+I^Q(\mathcal{V}(q),\mathcal{S}')) = s + I^Q(\mathcal{V}(p),\mathcal{S}')$, etc.

6) For each $p,q \in \bar{\mathbb{N}}^n$, $p \leq q$, the diagram is commutative:

$$\begin{array}{ccccccc} & & A^Q(\mathcal{V},\mathcal{S}',q) & \xrightarrow{\gamma_{q,p}} & A^Q(\mathcal{V},\mathcal{S}',p) & & \\ & & \uparrow \beta_q & & \uparrow \beta_p & & \\ \varepsilon_q & & \mathcal{V}(q) \oplus \mathcal{S}'/\mathcal{V}(q) & \xrightarrow{\eta_{q,p}} & \mathcal{V}(p) \oplus \mathcal{S}'/\mathcal{V}(p) & & \varepsilon_p \\ & & \downarrow \omega\circ\alpha_q & & \downarrow \omega\circ\alpha_p & & \\ & & \mathcal{D}'(\mathbb{R}^n) & \xleftrightarrow{\text{id}} & \mathcal{D}'(\mathbb{R}^n) & & \end{array}$$

with $\eta_{q,p}(s+\mathcal{V}(q)) = s + \mathcal{V}(p)$

therefore, $\gamma_{q,p}$ restricted to $\varepsilon_q(\mathcal{D}'(\mathbb{R}^n))$ is injective.

Proof

4) results from (20.2). The rest follows from Theorem 1 $\nabla\nabla\nabla$

The existence of derivative operators on the algebras as well as their properties are established now.

Theorem 3

In the case of derivative algebras (see § 6), suppose given $(V,\mathcal{S}') \in \mathcal{R}(P)$ and an admissible property Q, such that $Q \leq P$. Then

1) For each $p \in N^n$ and $q \in \bar{N}^n$, the following linear mapping (p-th order derivative) exists (see Remark D, § 7):

(26) $$D^p_{q+p} : A^Q(V,\mathcal{S}',q+p) \rightarrow A^Q(V,\mathcal{S}',q)$$

with

(26.1) $$D^p_{q+p}(s+I^Q(V(q+p),\mathcal{S}')) = D^p s + I^Q(V(q),\mathcal{S}')$$

and the restriction of D^p_{q+p} to $\mathcal{C}^\infty(R^n)$ is the usual p-th order derivative of functions.

2) The relation holds

(27) $$D^{p_1}_{q+p_1+p_2} \; D^{p_2}_{q+p_2} = D^{p_1+p_2}_{q+p_1+p_2} \; , \quad \forall \; p_1,p_2 \in N^n \; , \quad q \in \bar{N}^n$$

3) For each $p \in N^n$, $q,r \in \bar{N}^n$, $q \leq r$, the diagram is commutative:

$$\begin{array}{ccc} A^Q(V,\mathcal{S}',r+p) & \xrightarrow{\;D^p_{r+p}\;} & A^Q(V,\mathcal{S}',r) \\ \big\downarrow \gamma_{r+p,q+p} & & \big\downarrow \gamma_{r,q} \\ A^Q(V,\mathcal{S}',q+p) & \xrightarrow[\;D^p_{q+p}\;]{} & A^Q(V,\mathcal{S}',q) \end{array}$$

4) The mapping D^p_{q+p}, with $p \in N^n$, $q \in \bar{N}^n$, satisfies the Leibnitz rule of product derivative:

(28) $$D^p_{q+p}(S \cdot T) = \sum_{\substack{k \in N^n \\ k \leq p}} \binom{p}{k} \gamma_{q+p-k,q} \; D^k_{q+p} \; S \cdot \gamma_{q+k,q} \; D^{p-k}_{q+p} \; T \; ,$$

in particular, if $|p| = 1$, the relation holds:

(28.1) $$D^p_{q+p}(S \cdot T) = D^p_{q+p} \; S \cdot \gamma_{q+p,q} \; T + \gamma_{q+p,q} \; S \cdot D^p_{q+p} \; T \; ,$$

where $S,T \in A^Q(V,\mathcal{S}',q+p)$ in both of the above relations.

Proof

1) First, we prove (26). Obviously

(29) $D^p \mathcal{V}(q+p) \subset \mathcal{V}(q)$, $\forall\, p \in N^n$, $q \in \bar{N}^n$.

Now, we show that

(30) $D^p A^Q(\mathcal{V}(q+p),\mathcal{S}') \subset A^Q(\mathcal{V}(q),\mathcal{S}')$, $\forall\, p \in N^n$, $q \in \bar{N}^n$.

Indeed, (18) results in

$$D^p A^Q(\mathcal{V}(q+p),\mathcal{S}') \subset \cap\, D^p A$$

where the intersection is taken over all subalgerbras A in $\mathcal{W}$ which have the property Q and contain $\mathcal{V}(q+p) \oplus \mathcal{S}'$. Since we are in the presence of derivative algebras, each of the subalgebras A above satisfies the condition $D^p A \subset A$, therefore, one obtains

$$D^p A^Q(\mathcal{V}(q+p),\mathcal{S}') \subset \cap\, A$$

with the intersection having the same range as before. Noticing that $\mathcal{V}(q+p) \subset \mathcal{V}(q)$, one obtains

$$D^p A^Q(\mathcal{V}(q+p),\mathcal{S}') \subset \cap\, A$$

here the intersection being taken over all subalgebras A in $\mathcal{W}$ which have the property Q and contain $\mathcal{V}(q) \oplus \mathcal{S}'$. Taking now into account the definition in (18), one obtains (30).

The relations

(31) $A^Q(\mathcal{V}(q+p),\mathcal{S}') \subset A^Q(\mathcal{V}(q),\mathcal{S}')$, $\forall\, p \in N^n$, $q \in \bar{N}^n$

result easily noticing that $\mathcal{V}(q+p) \subset \mathcal{V}(q)$.

Comparing (29), (30) and (31), it follows that

(32) $D^p I^Q(\mathcal{V}(q+p),\mathcal{S}') \subset I^Q(\mathcal{V}(q),\mathcal{S}')$, $\forall\, p \in N^n$, $q \in \bar{N}^n$.

Now, (31) and (32) will imply (26).

The second part of 1) results from (20.2).

2), 3) and 4) result from 1) and Theorems 1,2 ∇∇∇

Now, positive powers will be defined for certain elements of the algebras. Denote

$$\mathcal{C}^\infty_+(R^n) = \left\{ \psi \in \mathcal{C}^\infty(R^n) \;\middle|\; \begin{array}{ll} *) & \psi(x) \geq 0\ ,\ \forall\, x \in R^n \\ **) & \psi^\alpha \in \mathcal{C}^\infty(R^n)\ ,\ \forall\, \alpha \in (0,\infty) \end{array} \right\}$$

Obviously, if $\psi \in \mathcal{C}^\infty(R^n)$ and $\psi(x) > 0$, $\forall\, x \in R^n$, then $\psi \in \mathcal{C}^\infty_+(R^n)$. But there exist $\psi \in \mathcal{C}^\infty(R^n)$, $\psi(x) \geq 0$, $\forall\, x \in R^n$, such that $\psi \notin \mathcal{C}^\infty_+(R^n)$, for instance, $\psi(x) = x_1^2 \ldots x_n^2$, $\forall\, x = (x_1, \ldots, x_n) \in R^n$. However, defining

$$\psi(x) = \begin{cases} \exp\left(-(1/x_1 + \ldots + 1/x_n)\right) & \text{if } x_i > 0 , \ \forall\, 1 \le i \le n \\ 0 & \text{otherwise} \end{cases}$$

one obtains $\psi \in \mathcal{C}_+(R^n)$. Denote further

$$W_+ = \{ s \in W \mid s(\nu) \in \mathcal{C}_+^\infty(R^n) , \quad \forall\, \nu \in N \}$$

It follows that, for $s \in W_+$ one can define any positive power s^α , by

(33) $$s^\alpha(\nu)(x) = (s(\nu)(x))^\alpha , \quad \forall\, \alpha \in (0,\infty) , \quad \nu \in N , \quad x \in R^n ,$$

and one obtains $s^\alpha \in W_+$.

Now, the condition in (16) for a positive power invariant subset H in W can be reformulated as follows

(34) $$\forall\ s \in H \cap W_+ , \quad \alpha \in (0,\infty) : s^\alpha \in H$$

Suppose T is a vector subspace in S_0 and denote

$$D'_{T,+}(R^n) = \{ \langle t,\cdot\rangle \mid t \in T \cap W_+ \}$$

The distributions in $D'_{T,+}(R^n)$ will be called <u>T-nonnegative.</u>

<u>Theorem 4</u>

In the case of <u>positive power algebras</u>, suppose given $(V,S') \in R(P)$ and an admissible porperty Q , such that $Q \le P$. If T is a vector subspace in S_0 , satisfying the condition

(35) $$U \cap W_+ \subset T \subset S'$$

then

1) $\mathcal{C}_+^\infty(R^n) \subset D'_{T,+}(R^n)$

2) For given $p \in \bar{N}^n$ and any $\alpha \in (0,\infty)$, one can define a mapping (positive power):

$$D'_{T,+}(R^n) \ni T \longrightarrow T^\alpha \in A^Q(V,S',p)$$

with $T = \langle t,\cdot\rangle \to T^\alpha = t^\alpha + I^Q(V(p),S')$, where $t \in T \cap W_+$ and the relati ons will result:

(36.1) $$T^1 = T$$

(36.2) $$T^{\alpha+\beta} = T^\alpha \cdot T^\beta , \quad \forall\, \alpha,\beta \in (0,\infty)$$

(36.3) $$(T^\alpha)^m = T^{\alpha \cdot m} , \quad \forall\, \alpha \in (0,\infty) , \quad m \in N\setminus\{0\} .$$

3) For any $p \in \bar{N}^n$, the mapping in 1) is identical on $\mathcal{C}_+^\infty(R^n)$ with the usual positive power of functions.

4) Suppose in addition the case of <u>derivative algebras</u>, then the following relations hold in the algebras $A^Q(V,S',p)$, with $p \in \bar{N}^n$:

(37) $$D^q_{p+q} T^\alpha = \alpha \cdot T^{-1} D^q_{p+q} T ,$$

$$\forall \quad T \in D'_{T,+}(R^n) , \quad \alpha \in (1,\infty) , \quad q \in N^n , \quad |q| = 1$$

Proof

1) It follows from (35). 2) It results from (35) and 2) in Theorem 2. 3) It results from 1) and 2). Finally, 4) is a consequence of (28.1) ∇∇∇

Remark 2

The condition (35) can be easily fulfilled in case (20.2) is replaced by the stronger condition (see (25) in chap. 2, §6): $U \subset S'$.

§9. DEFINING NONLINEAR PARTIAL DIFFERENTIAL OPERATORS ON THE ALGEBRAS

As stated in §1, one of the main aims of the nonlinear method in the theory of distributions presented in this work is to offer a framework for the study of the nonlinear partial differential equations. The embedding of the distributions in $D'(R^n)$ into the algebras $A^Q(V,S',p)$ (see Theorem 2, §8) creates the possibility of studying <u>nonlinear partial differential operators</u> of the general form

$$T(D)u(x) = \sum_{1 \le i \le h} c_i(x) \prod_{1 \le j \le k_i} D^{p_{ij}} u(x) , \quad x \in R^n ,$$

$$(\text{or} \quad x \in \Omega , \quad \Omega \subset R^n , \quad \Omega \neq \emptyset , \quad \text{open})$$

with c_i smooth and $p_{ij} \in N^n$.

In order to define these operators on the algebras containing the distributions, one has to take into account the following two features of the distribution multiplication presented in this work:

1) The algebra homomorphisms (see 5) in Theorem 2, §8)

$$A^Q(V,S',q) \xrightarrow[\gamma_{q,p}]{} A^Q(V,S',p)$$

need not generate algebra embeddings.

2) The derivative operators (see pct. 1 in Theorem 3, §8)

$$D^p_{q+p} : A^Q(V,S',q+p) \to A^Q(V,S',q)$$

act between different algebras in case $q \in \bar{N}^n$ has finite components.

Suppose, we are in the case of <u>derivative algebras</u>. Given the above operator $T(D)$, define its order by

$$\bar{p} = \max \{ p_{ij} \mid 1 \le i \le h \ , \ 1 \le j \le k_i \}$$

then, one can define for any $q \in \bar{N}^n$ the mapping

$$T(D) : A^Q(V, S', q+\bar{p}) \rightarrow A^Q(V, S', q)$$

by

$$T(D)u = \sum_{1 \le i \le h} c_i \prod_{1 \le j \le k_i} \gamma_{q+\bar{p}-p_{ij}, q} D_{q+\bar{p}}^{p_{ij}} u \ , \quad \forall \ u \in A^Q(V, S', q+\bar{p}) \ .$$

where the additions and multiplications in the right term are effectuated in $A^Q(V, S', q)$. The commutativity of the diagrams in 5) in Theorem 2 and 3) in Theorem 3, §8, grants the consistency of the above definition. Moreover, due to 4) in Theorem 2, §3, the above definition of $T(D)$ coincides with the usual one in the case of smooth u .

An important particular case is when $T(D)$ acts upon $u \in D'(R^n)$. Then, due to 3) in Theorem 2, §3, one can consider u belonging to any of the algebras $A^Q(V, S', r)$, with $r \in \bar{N}^n$, $r \ge \bar{p}$, and apply the above definition.

Due to 5) in Theorem 2,§3, the same happens when $T(D)$ acts upon $u \in A^Q(V, S', \infty)$.

In chapter 3, the above definition will be used in the study of piece wise smooth weak solutions of the so called polynomial nonlinear partial differential operators. The nonlinear hyperbolic partial differential equations modelling the shock waves are particular cases of polynomial nonlinear partial differential equations.

§10. MAXIMALITY AND LOCAL VANISHING

There exists an applicative interest in constructing the algebras $A^Q(V, S', p)$, $p \in \bar{N}^n$ in a way that the ideals $I^Q(V(p), S')$, $p \in \bar{N}^n$, are <u>large</u>, possibly <u>maximal</u>. Indeed, according to (24), the larger these ideals, the more equality relations are obtained in the corresponding algebras.

To construct the algebras $A^Q(V, S', p)$ and the ideals $I^Q(V(p), S')$ means to choose P-regularizations (V, S') . According to (11) and (19), the larger V is, the larger the ideals $I^Q(V(p), S')$, $p \in \bar{N}^n$, will be. Therefore, as a first approach in securing maximal ideals, the problem of $(V, S') \in R(P)$, with maximal V , will be studied in the present section. An alternative approach to the problem of maximal ideals $I^Q(V(p), S')$, $p \in \bar{N}^n$, will be given in chap. 2, §7.

And now, several results on the structure of $R(P)$. In addition to the relation

(38) $\quad \forall \ (V, S') \in R(P) : V \subsetneqq V_0$

obtained in 2), Remark 1, §7, the following two simple results will be useful.

Lemma 2

Suppose $(V,S') \in R(P)$. If S'' is a vector subspace in S_o satisfying the conditions

(39) $$V \oplus S'' = V \oplus S'$$

(40) $$U \subset V(p) \oplus S'' , \quad \forall \; p \in \bar{N}^n \quad \text{(see (20.2))}$$

then $(V,S'') \in R(P)$.

Proof

Since $V \subset V_o$, the relations (39) and (20.1) will obviously imply $S_o = V_o \oplus S''$. Now, it suffices to show that (V,S'') satisfies (20.3). Indeed, taking into account (19) and (18), one obtains $I^P(V,S'') = I^P(V,S')$ ▽▽▽

Lemma 3

Suppose $(V_1 , S_1') , (V_2 , S_2') \in R(P)$. If $S_1' \subset S_2'$ then $S_1' = S_2'$.

Proof

It follows easily from the fact that both (V_1 , S_1') and (V_2 , S_2') satisfy (20.1) ▽▽▽

The relations (11), (19) and Lemmas 2,3 suggest that an appropriate way in enlarging the ideals $I^Q(V(p),S')$, $p \in \bar{N}^n$, is to enlarge V . From here the interest in finding $(V,S') \in R(P)$ with V maximal.

Define on $R(P)$ a partial order $\leq$ by

$$(V_1 , S_1') \leq (V_2 , S_2') \iff V_1 \subset V_2 \text{ and } S_1' = S_2'$$

The admissible property P is called regular, only if each chain in $(R(P) , \leq)$ has an upper bound.

We recall (see §6) that the strongest admissible property, namely the property of subsets H in W that $H = W$, was denoted by $\bar{P}$.

Theorem 5

$\bar{P}$ is a regular admissible property.

Proof

Assume $((V_\lambda , S') \mid \lambda \in \Lambda)$ is a chain in $(R(P) , \leq)$. Denote $V = \bigcup_{\lambda \in \Lambda} V_\lambda$. We shall prove that $(V,S') \in R(P)$.

Obviously, V is a vector subspace in V_0 and (20.1) and (20.2) hold. It remains to prove (20.3). First, we notice that $A^{\bar{P}}(V,S') = W$, since $\bar{P}$ holds only for $A = W$. Therefore, $I^P(V,S') = I(V,W)$ and the condition (20.3) becomes $I(V,W) \cap S' = O$. Assume now $s \in I(V,W) \cap S'$. Then, $s \in I(V,W)$ results in

$$(41) \qquad s = \sum_{1\le i\le k} v_i \cdot w_i\,, \quad \text{with} \quad v_i \in V\,, \quad w_i \in W\,, \quad \forall\ 1\le i\le k\,.$$

Since $((V_\lambda, S') \mid \lambda \in \Lambda)$ is a chain, there exists $\lambda_0 \in \Lambda$ such that $v_i \in V_{\lambda_0}$, $\forall\ 1\le i\le k$. Now, (41) will imply

$$s \in I(V_{\lambda_0}, W) \cap S' = O$$

the last equality resulting from the fact that $(V_{\lambda_0}, S') \in R(P)$ ∇∇∇

Based on Zorn's Lemma, Theorem 1, §7 and Theorem 5, one can define for a given regular admissible property P the nonvoid set of maximal P-regularizations:

$$R_{max}(P) = \{ (V,S') \in R(P) \mid (V,S') \text{ maximal in } (R(P), \le) \}$$

The fact that the strongest admissible property $\bar{P}$ is regular (see Theorem 5) offers the possibility to construct the algebras $A^Q(V,S',p)$, $p \in \bar{N}^n$, for any admissible property Q, with V maximal. Indeed, since $Q \le \bar{P}$, one can choose $(V,S') \in R_{max}(\bar{P})$ and the construction of the algebras will proceed according to (24) and (23).

The above remark and the relation (38) generate an interest in a possible upper bound of V, with $(V,R') \in R_{max}(\bar{P})$, which would give an insight into the necessary structure of the distribution multiplications defined according to the relations (24), (23), (18), (19) and (11).

The following Theorem 6 gives an upper bound of the mentioned type under the form of a local vanishing property which V has to satisfy. Namely, it is proved for instance, that in case $V \oplus S'$ contains 'δ sequences' (see Lemma 1 in §4 and [4], [35-41], [53], [68], [69], [105-110], [136], [137], [162]) the sequences of smooth functions weakly convergent to $0 \in D'(R^n)$ which constitute V, have to vanish infinitely many times in points arbitrarily near to each point in R^n.

For $p \in \bar{N}^n$, denote by W^p the set of all sequences of smooth functions $w \in W$ which satisfy the local vanishing property:

$$(42) \qquad \begin{array}{l} \forall\ G \subset R^n,\ G \ne \emptyset,\ \text{open},\ q \in N^n,\ q \le p,\ \mu \in N: \\ \exists\ x \in G,\ \nu \in N,\ \nu \ge \mu: \\ \quad D^q w(\nu)(x) = 0 \end{array}$$

or, written simpler

$$(42') \qquad \begin{array}{l} \forall\ x \in R^n,\ q \in N^n,\ q \le p: \\ \quad D^q w(\nu)(y) = 0 \ \text{ for infinitely many } \nu \in N \\ \qquad\qquad\qquad \text{and } y \in R^n \text{ arbitrarily near to } x \end{array}$$

A P-regularization (V,S') is of <u>local type</u>, only if

(43) $\forall\ G \subset R^n\ ,\ G \neq \emptyset\ ,\ \text{open}\ :$
$\exists\ s_G \in V + S'\ :$

(43.1) $\langle s_G\ ,\ \cdot\rangle \neq 0 \in D'(R^n)$

(43.2) $\text{supp}\ s_G(\nu) \subset G\ ,\quad \forall\ \nu \in N\ ,\ \nu \geq \mu\ ,$

for a certain $\mu \in N$.

<u>Theorem 6</u>

Suppose given a local type regularization $(V,S') \in R(\bar{P})$ such that $V \oplus S'$ is sectional invariant.

Then

$$V(p) \subset W^p\ ,\quad \forall\ p \in \bar{N}^n\ .$$

<u>Proof</u>

Taking into account (22) and (42), it suffices to prove the inclusion $V \subset W^0$ only. Assume it is false and $v \in V \setminus W^0$. Then (42) implies

$$\exists\ G \subset R^n\ ,\ G \neq \emptyset\ ,\ \text{open}\ ,\ \mu' \in N :$$
$$\forall\ x \in G\ ,\quad \nu \in N\ ,\ \nu \geq \mu'\ :$$
$$v(\nu)(x) \neq 0$$

Now, due to (43.2), one obtains

$$\exists\ \mu'' \in N :$$
$$\forall\quad \nu \in N\ ,\ \nu \geq \mu''\ :$$
$$\text{supp}\ s_G(\nu) \subset G$$

Define $w \in W$ by

$$w(\nu)(x) = \begin{cases} 0 & \text{if } x \notin G \\ s_G(\nu)(x)/v(\nu)(x) & \text{if } x \in G \end{cases}$$

whenever $\nu \in N\ ,\ \nu \geq \mu = \max\{\mu'\ ,\ \mu''\}$. Then

(44) $v(\nu)\cdot w(\nu) = s_G(\nu)\ ,\quad \forall\ \nu \in N\ ,\ \nu \geq \mu$

therefore

(45) $v \cdot w \in V \oplus S'$

since $s_G \in V \oplus S'$ and $V \oplus S'$ is sectional invariant. But

(46) $v \cdot w \in I(V,W) = I^{\bar{P}}(V,S')$

since $v \in V$.

Now, (45) and (46) together with (20.3) will imply $v \cdot w \in V$ which due to (44) results in $\langle s_G , \cdot \rangle = 0 \in \mathcal{D}'(R^n)$, since $V \subset V_0$. Therefore (43.1) was contradicted. ∇∇∇

Remark 3

There exist relevant instances of regularizations (V,S') which meet the conditions in Theorem 6 above. Indeed, one can notice that $V \oplus S'$ will be sectional invariant whenever V has that property. Further, the regularizations $(V, T \oplus S_1)$ obtained in Theorem 4, chap. 2, §6, can be chosen with V sectional invariant, since any Dirac ideal obtained in Proposition 6, chap. 2, §6 is obviously sectional invariant. Finally, the regularizations $(V, T \oplus S_1)$ considered in chap. 5, §4, are of local type. Indeed, $T_\Sigma \subset T$ where $\Sigma = (s_x \mid x \in R^n) \in Z_\delta$, therefore given $G \subset R^n$, $G \neq \emptyset$, open, one can take in (43) $s_G = s_x$, provided that $x \in G$.

With the method used in the proof of Theorem 6 one obtains the following more general result.

Theorem 7

Suppose given a regularization (V,S') such that $V \oplus S'$ is sectional invariant.

Then each $v \in V$ satisfies the vanishing condition

$$\forall \; t \in V \oplus S' , \; t \notin V , \; \mu \in N ;$$
$$\exists \; \nu \in N , \; \nu \geq \mu , \; x \in \operatorname{supp} t(\nu) :$$
$$v(\nu)(x) = 0$$

Proof

Assume it is false and $v \in V$, $t \in V \oplus S'$, $t \notin V$ and $\mu \in N$ such that

$$\forall \; \nu \in N , \; \nu \geq \mu :$$
$$v(\nu) \neq 0 \text{ on } \operatorname{supp} t(\nu)$$

Define $w \in W$ by

$$w(\nu)(x) = \begin{cases} 0 & \text{if } x \notin \operatorname{supp} t(\nu) \\ t(\nu)(x)/v(\nu)(x) & \text{if } x \in \operatorname{supp} t(\nu) \end{cases}$$

whenever $\nu \in N$, $\nu \geq \mu$. Then

$$(47) \qquad v(\nu) \cdot w(\nu) = t(\nu) , \quad \forall \; \nu \in N , \; \nu \geq \mu$$

therefore

$$(48) \qquad v \cdot w \in V \oplus S'$$

since $t \in V \oplus S'$ and $V \oplus S'$ is sectional invariant.

But

(49) $\quad v \cdot w \in I(V,W) = I^{\bar{P}}(V,S')$

since $v \in V$.

The relations (48) and (49) together with (20.3) imply $v \cdot w \in V$. Then (47) will give $\langle t,\cdot\rangle = 0 \in D'(R^n)$ since $V \subset V_0$. It follows that $t \in V_0$. Now, (20.1) will contradict $t \notin V$ $\nabla\nabla\nabla$

In the same way, one can prove:

<u>Theorem 8</u>

Suppose (V,S') is a regularization, then each $v \in V$ satisfies the <u>vanishing</u> condition

$$\begin{aligned} &\forall \; t \in V \oplus S' \;, \quad t \notin V : \\ &\exists \; \nu \in N \;, \quad x \in \text{supp}\; t(\nu) : \\ &\quad v(\nu)(x) = 0 \end{aligned}$$

§11. STRONGER CONDITIONS FOR DERIVATIVES

It will be shown that even in the one dimensional case $n = 1$, the stronger conditions on derivatives mentioned in Remark D, §7, lead necessarily to a particular, rather trivial distribution multiplication.

Suppose A is an associative and commutative algebra containing the real valued polynomials on R^1 as well as the distributions in $D'(R^1)$ with support a finite number of points.

Suppose also that

(50) the multiplication in A induces the usual multiplication on the polynomials and the polynomial $\psi(x) = 1$, $\forall \; x \in R^1$, is the unit element in A ,

(51) there exists a linear mapping $D : A \to A$ such that

(51.1) D is identical with the usual derivative when applied to polynomials or distributions with support a finite number of points

(51.2) D satisfies on A the Leibnitz rule of 'product derivative'

$D(a\cdot b) = (Da) \cdot b + a \cdot (Db) \;, \quad \forall \; a,b \in A$

and finally

(52) $\quad (x-x_0) \cdot \delta_{x_0} = 0 \in A \;, \quad \forall \; x_0 \in R^1$

Theorem 9

Within the algebra A the relations hold:

(53) $$(x-x_0)^p \cdot D^q\delta_{x_0} = 0 \in A\ , \quad \forall\ x_0 \in R^1\ , \quad p,q \in N\ , \quad p > q$$

(54) $$(p+1) \cdot D^p\delta_{x_0} + (x-x_0) \cdot D^{p+1}\delta_{x_0} = 0 \in A\ , \quad \forall\ x_0 \in R^1\ , \quad p \in N$$

(55) $$(x-x_0)^p \cdot (D^p\delta_{x_0})^q = 0 \in A\ , \quad \forall\ x_0 \in R^1\ , \quad p,q \in N\ , \quad q \geq 2\ .$$

(56) $$(\delta_{x_0})^2 = \delta_{x_0} \cdot D\delta_{x_0} = 0 \in A\ , \quad \forall\ x_0 \in R^1$$

Proof

Applying D to (52) and taking into account (51), one obtains

(57) $$\delta_{x_0} + (x-x_0) \cdot D\delta_{x_0} = 0 \in A\ , \quad \forall\ x_0 \in R^1$$

which multiplied by $(x-x_0)$ gives due to (52) the relation $(x-x_0)^2 \cdot D\delta_{x_0} = 0 \in A$, $\forall\ x_0 \in R^1$. Applying D to the latter relation and then, multiplying by $(x-x_0)$, one obtains in the same way the relation $(x-x_0)^3 \cdot D^2\delta_{x_0} = 0 \in A$, $\forall\ x_0 \in R^1$. Repeating the procedure, one obtains (53).

The relation (54) results applying repeatedly D to (57).

Now, multiplying (54) by $(x-x_0)^p$, one obtains

$$(p+1)(x-x_0)^p \cdot D^p\delta_{x_0} + (x-x_0)^{p+1} \cdot D^{p+1}\delta_{x_0} = 0 \in A\ , \quad \forall\ x_0 \in R^1\ , \ p \in N$$

Multiplying that relation by $(D^p\delta_{x_0})^{q-1}$ and taking into account (53), one obtains (55).

Taking $p = 0$ and $q = 2$ in (55), one obtains $(\delta_{x_0})^2 = 0 \in A$, $\forall\ x_0 \in R^1$. Applying D to that relation, the proof of (56) is completed ▽▽▽

§12. APPENDIX

The proof of Lemma 1 in §4 is given here.

1) It follows easily.

2) For $a \in R^1$ and $\nu \in N$ denote

$$E(a,\nu) = \{\ x \in R^n \mid s(\nu)(x) \geq a\ \}$$

First, we prove the relation

(58) $$\overline{\lim_{\nu\to\infty}} \int_{E(a,\nu)} s(\nu)(x)dx \geq 1 , \quad \forall \quad a \in R^1$$

Assume it is false. Then

$$\exists \quad a \in R^1 , \quad \varepsilon > 0 , \quad \mu' \in N :$$
$$\forall \quad \nu \in N , \quad \nu \geq \mu' :$$
$$\int_{E(a,\nu)} s(\nu)(x)dx \leq 1 - \varepsilon$$

But, $s \in S_o$, $\langle s,\cdot\rangle = \delta$ and supp $s(\nu)$ shrinks to $0 \in R^n$, when $\nu \to \infty$. Therefore, assuming $\psi \in D(R^n)$ and $\psi = 1$ on a neighbourhood of $0 \in R^n$, one obtains

$$1 = \psi(0) = \lim_{\nu\to\infty} \int_{R^n} s(\nu)(x)\psi(x)dx = \lim_{\nu\to\infty} \int_{R^n} s(\nu)(x)dx$$

It follows that

$$\exists \quad \mu'' \in N :$$
$$\forall \quad \nu \in N , \quad \nu \geq \mu'' :$$
$$1 - \varepsilon/2 \leq \int_{R^n} s(\nu)(x)dx$$

Now, for $\nu \in N$, the relations hold

$$\int_{R^n} s(\nu)(x)dx = \int_{E(a,\nu)} s(\nu)(x)dx + \int_{\text{supp } s(\nu)\setminus E(a,\nu)} s(\nu)(x)dx \leq$$
$$\leq \int_{E(a,\nu)} s(\nu)(x)dx + a \int_{\text{supp } s(\nu)} dx$$

Therefore, one obtains for $\nu \in N$, $\nu \geq \max \{\mu' , \mu''\}$ the inequality

$$1 - \varepsilon/2 \leq 1 - \varepsilon + a \int_{\text{supp } s(\nu)} dx$$

which is absurd since supp $s(\nu)$ shrinks to $0 \in R^n$, and the proof of (58) is completed.

We prove now that there exist $a_\nu \in [0,\infty)$, with $\nu \in N$, such that

(59) $$\lim_{\nu\to\infty} a_\nu = \infty \quad \text{and} \quad \overline{\lim_{\nu\to\infty}} \int_{E(a_\nu,\nu)} s(\nu)(x)dx \geq 1$$

Indeed, according to (58), there exist $\nu_\mu \in N$, with $\mu \in N$ such that

(60) $$\nu_0 < \nu_1 < \ldots < \nu_\mu < \ldots$$

and

(61) $$1 - 1/(\mu+1) \leq \int_{E(\mu,\nu_\mu)} s(\nu_\mu)(x)dx , \quad \forall \ \mu \in N$$

Define now $a_\nu = \inf \{ \mu \in N \mid \nu \leq \nu_\mu \}$, with $\nu \in N$.
Then $a_\nu \leq a_{\nu+1}$, $\forall \ \nu \in N$ and

(62) $$a_{\nu_\mu} = \mu , \quad \forall \ \mu \in N$$

due to (60), hence, the first relation in (59) is proved. Taking into account (61), the second relation in (59) follows from (62).

Finally, we prove

(63) $$\overline{\lim_{\nu\to\infty}} \int_{E(a_\nu,\nu)} (s(\nu)(x))^2 dx = +\infty$$

Indeed, $(s(\nu))^2 \geq a_\nu s(\nu)$ on $E(a_\nu,\nu)$, $\forall \ \nu \in N$, since $s(\nu) \geq a_\nu \geq 0$ on $E(a_\nu,\nu)$ $\forall \ \nu \in N$. Therefore

$$\int_{E(a_\nu,\nu)} (s(\nu)(x))^2 dx \geq a_\nu \cdot \int_{E(a_\nu,\nu)} s(\nu)(x)dx , \quad \forall \ \nu \in N$$

The relation (63) will result now from (59). Obviously, (59) implies

$$\overline{\lim_{\nu\to\infty}} \int_{R^n} (s(\nu)(x))^2 dx = +\infty$$

Then $s^2 \notin \mathcal{S}_0$ since $\operatorname{supp} s^2(\nu) = \operatorname{supp} s(\nu)$ shrinks to $0 \in R^n$ when $\nu \to \infty$.

Remark 4

The condition of nonnegativity of the sequence s in 1) in Lemma 1, §4, can be removed in special cases. For instance, assume s given by

$$s(\nu)(x) = a_\nu \psi(b_\nu x) , \quad \forall \nu \in N , \quad x \in R^n ,$$

where $\psi \in D(R^n)$, $a_\nu \in C^1$, $b_\nu \in R^1$ and $\lim_{\nu\to\infty} | b_\nu | = + \infty$. Then, it is easy to see that the equivalence between 1.1) and 1.2) in the mentioned lemma, will still be valid.

Chapter 2

DIRAC ALGEBRAS CONTAINING THE DISTRIBUTIONS

§1. INTRODUCTION

In chapter 1, diagrams of inclusions of the general type (23) were constructed in order to obtain the algebras (24) containing the distributions in $\mathcal{D}'(\mathbb{R}^n)$. The construction of diagrams (23) was based on the presumed existence (Theorem 1, chap. 1, §7) of P-regularizations (V,S') , for a given admissible property P .

In this chapter two results are presented.
First, specific instances of the diagrams (23), chap. 1, §7, are constructed, leading to so called Dirac algebras in which nonlinear operations of polynomial type can be performed with piece wise smooth functions on $\mathbb{R}^n$ and their distributional derivatives. The nonlinear operations considered, cover the ones encountered in the nonlinear partial differential operators introduced in chap. 1, §9. In that way, the Dirac algebras prove to be useful in chapter 3, in the study of nonlinear partial differential equations with piece wise smooth weak solutions. The class of the piece wise smooth functions admitted in the nonlinear operations is rather wide, their singularities being situated on arbitrary closed subsets of $\mathbb{R}^n$ with smooth boundaries, for instance, locally finite families of smooth surfaces in $\mathbb{R}^n$.

As a second result, based on the existence of Dirac algebras, one can prove the existence of the regularizations (V,S') used in chapter 1, and therefore validate the general method of embedding the distributions into algebras, presented there. For an alternative validation, not using Dirac algebras, see §§8 and 9.

§2. CLASSES OF SINGULARITIES OF PIECE WISE SMOOTH FUNCTIONS

When performing nonlinear operations with piece wise smooth functions on $\mathbb{R}^n$ and their distributional derivatives, a problem arises in the neighbourhood of the singularities. The classes of singularities, concentrated on arbitrary closed subsets of $\mathbb{R}^n$ with smooth boundaries, for instance, locally finite families of smooth surfaces in $\mathbb{R}^n$, are defined now.

A set Γ of mappings $\gamma : \mathbb{R}^n \to \mathbb{R}^{m_\gamma}$, $\gamma \in \mathcal{C}^\infty$, with $m_\gamma \in \mathbb{N}$, is called a singularity generator on $\mathbb{R}^n$. The closed subsets in $\mathbb{R}^n$

$$F_\gamma = \{ x \in R^n \mid \gamma(x) = 0 \in R^{m_\gamma} \}$$

defined by the mappings $\gamma \in \Gamma$ will represent the basic sets of possible singularities

The set F_Γ of all $F_\Delta = \bigcup_{\gamma \in \Delta} F_\gamma$, where $\Delta \subset \Gamma$ and F_Δ is closed, will be called the <u>class of singularities associated to Γ</u>. Obviously, if $\Delta \subset \Gamma$ and Δ is finite or more generally, $(F_\gamma \mid \gamma \in \Delta)$ is locally finite in R^n, then $F_\Delta \in F_\Gamma$. Therefore, we shall in the sequel be able to consider singularities concentrated on arbitrary locally finite families of smooth surfaces in R^n.

Denote then by $F_{\Gamma,loc}$ the set of all F_Δ with $\Delta \subset \Gamma$ and $(F_\gamma \mid \gamma \in \Delta)$ locally finite in R^n. It follows that $F_{\Gamma,loc} \subset F_\Gamma$.

<u>Remark 1</u>

The subsets F_γ can be fairly complicated. For instance, suppose $m_\gamma = 1$ and $\gamma(x_1, \ldots, x_n) = \exp(-1/x_1^2) \sin(1/x_1)$ if $x_1 \neq 0$, while $\gamma(x) = 0$ otherwise. Then F_γ is an infinite set of hyperplanes in R^n which is <u>not</u> locally finite. However, obviously $F_\gamma \in F_{\Gamma,loc}$.

The piece wise smooth functions on R^n considered will be those in

$$C^\infty_\Gamma(R^n) = \{ f : R^n \to C^1 \mid \exists \ F \in F_\Gamma : f \in C^\infty(R^n \setminus F) \}$$

thus, having the singularities concentrated on arbitrary closed subsets of R^n with smooth boundaries, for instance locally finite families of surfaces from Γ.

The nonlinear operations on functions in $C^\infty_\Gamma(R^n)$ and their distributional derivatives will be of the following polynomial type

$$(1) \qquad T(f_1, \ldots, f_m) = \sum_{1 \le i \le h} c_i \prod_{1 \le j \le k_i} D^{p_{ij}} g_{ij}$$

where $c_i \in C^\infty(R^n)$, $p_{ij} \in N^n$ and $g_{ij} \in \{f_1, \ldots, f_m\} \subset C^\infty_\Gamma(R^n)$.

The actual range of the nonlinear operations (1) will be the set of distributions

$$(C^\infty_\Gamma(R^n) \cap L^1_{loc}(R^n)) + D'_\Gamma(R^n)$$

where

$$D'_\Gamma(R^n) = \{ S \in D'(R^n) \mid \exists \ F \in F_\Gamma : \text{supp } S \subset F \}.$$

<u>§3. COMPATIBLE IDEALS AND VECTOR SUBSPACES OF SEQUENCES OF SMOOTH FUNCTIONS</u>

The construction of the Dirac algebras will proceed through §§3-6 in several stages, ending with Theorem 4 in §6.

Given a regularization (V,S') , one obtains (see Theorem 2, chap. 1, §8) the following embeddings of $D'(R^n)$ into algebras

$$
(2)\qquad
\begin{array}{ccccccc}
D'(R^n) & & S_0/V_0 & & V(p)\oplus S'/V(p) & & A^Q(V,S',p) \\
\Psi & & \Psi & & \Psi & & \Psi \\
\langle s,\cdot\rangle & \xleftarrow[\text{bij}]{} & s+V_0 & \xleftarrow[\text{bij}]{} & s+V(p) & \xrightarrow[\text{inj}]{} & s+I^Q(V(p),S')
\end{array}
$$

where Q is any admissible property and $p \in \bar{N}^n$.

It follows (see also Theorem 3, chap. 1, §8) that the nonlinear operations of type (1) when applied to distributions - in particular, functions in $C^\infty_\Gamma(R^n)$ - are effectuated within the algebras, according to the relation

$$
(3)\qquad
\begin{aligned}
& T(\langle s_1,\cdot\rangle, \dots, \langle s_m,\cdot\rangle) = \\
& = \sum_{1\le i\le h} c_i \prod_{1\le j\le k_i} D^{p_{ij}} s_{ij} + I^Q(V(p),S') \in A^Q(V,S',p)
\end{aligned}
$$

where $s_{ij} \in \{ s_1 ,\dots,s_m \} \subset V(p)\oplus S'$. One can always assume that $s_1 ,\dots,s_m \in S'$ in (3) since $V(p) \subset V_0$ and in the left term, the distributions $\langle s_1 ,\cdot\rangle,\dots,\langle s_m ,\cdot\rangle$ appear only. Therefore, S' has a particularly important role, since the nonlinear operations (1) and (3) when observed from S' become the corresponding classical operations applied term by term to sequences of smooth functions. The role V will have is to generate ideals $I^Q(V(p),S')$ which annihilate within the embeddings (2) the effect the singular distributions in $D'_\Gamma(R^n)$ cause in the nonlinear operations (1) and (3).

In this respect, the regularizations (V,S') will be chosen as follows:

a) V will be a vector subspace in $I \cap V_0$, where I is an ideal in W of sequences of smooth functions vanishing on certain singularities $F \in F_\Gamma$, as well as on neighbourhoods of points outside of those singularities.

b) S' will be split into $T\oplus S_1$, where the sequences of weakly convergent smooth functions in T represent the distributions in $D'_\Gamma(R^n)$.

The main part of the construction, both theoretical (in this chapter) and applicative (in chapters 3, 4 and 5) rests upon the ideals I .

The final choice of the ideals I and vector subspaces T and S_1 obtained in §6, will evolve in several steps.

It is particularly important to point out that the above way of choosing a regularization (V,S') belongs to a natural, general framework presented in Theorem 1 below, where a basic characterization of regularizations is given. That characterization will be used throughout the chapters 3-7, when constructing algebras containing the distributions needed in applications to nonlinear problems or in theoretical developments.

An ideal I in W and a vector subspace T in S_o are called compatible, only if (see Fig. 1.):

(4) $I \cap T = V_o \cap T = 0$

(5) $I \cap S_o \subset V_o \oplus T$

Theorem 1

Suppose the ideal I and vector subspace T are compatible. If V is a vector subspace in $I \cap V_o$ and S_1 is a vector subspace in S_o satisfying

(6) $V_o \oplus T \oplus S_1 = S_o$

(7) $U \subset V(p) \oplus T \oplus S_1$, $\forall\ p \in \bar{N}^n$

then $(V, T \oplus S_1) \in R(P)$ for any admissible property P . (see Fig. 2)

Conversely, any regularization (V,S') can be written under the above form.

Proof

Denote $S' = T \oplus S_1$. It suffices to show that (see (20.3) in chap. 1, §7)

(8) $I(V,W) \cap S' = 0$

First, we notice that $I(V,W) \subset I$ since $V \subset I$ and I is an ideal in W . Therefore

(9) $I(V,W) \cap S' \subset I \cap S'$

But

(10) $I \cap S' = 0$

Indeed, (5) results in

(11) $I \cap S' \subset (I \cap S_o) \cap S' \subset (V_o \oplus T) \cap (T \oplus S_1) \subset T$

the last inclusion being implied by (6). Now, (10) follows from (11) and (4). The relations (9) and (10) imply (8).

Conversely, assume given $(V,S') \in R(\bar{P})$ and denote $I = I^{\bar{P}}(V,S')$. Then, obviously $I = I(V,W)$ therefore, I is an ideal in W . But $V_o \oplus S' = S_o$, due to (20.1). Hence, there exists a vector subspace $T \subset S'$ such that $I \cap S_o \subset V_o \oplus T$. Obviously, one can choose a vector subspace S_1 in S' so that $S' = T \oplus S_1$. Now, (20.3) will imply $I \cap T \subset I \cap S' = 0$ while (20.2) will result in $U \subset V(p) \oplus T \oplus S_1$, $\forall\ p \in \bar{N}^n$ The proof is completed, noticing that $V \subset I(V,W) = I$, since $u(1) \in W$ ∇∇∇

Remark 2

Theorem 1 gives an affirmative answer to the question of the existence of regularizations (V,S') provided one can prove the existence of:

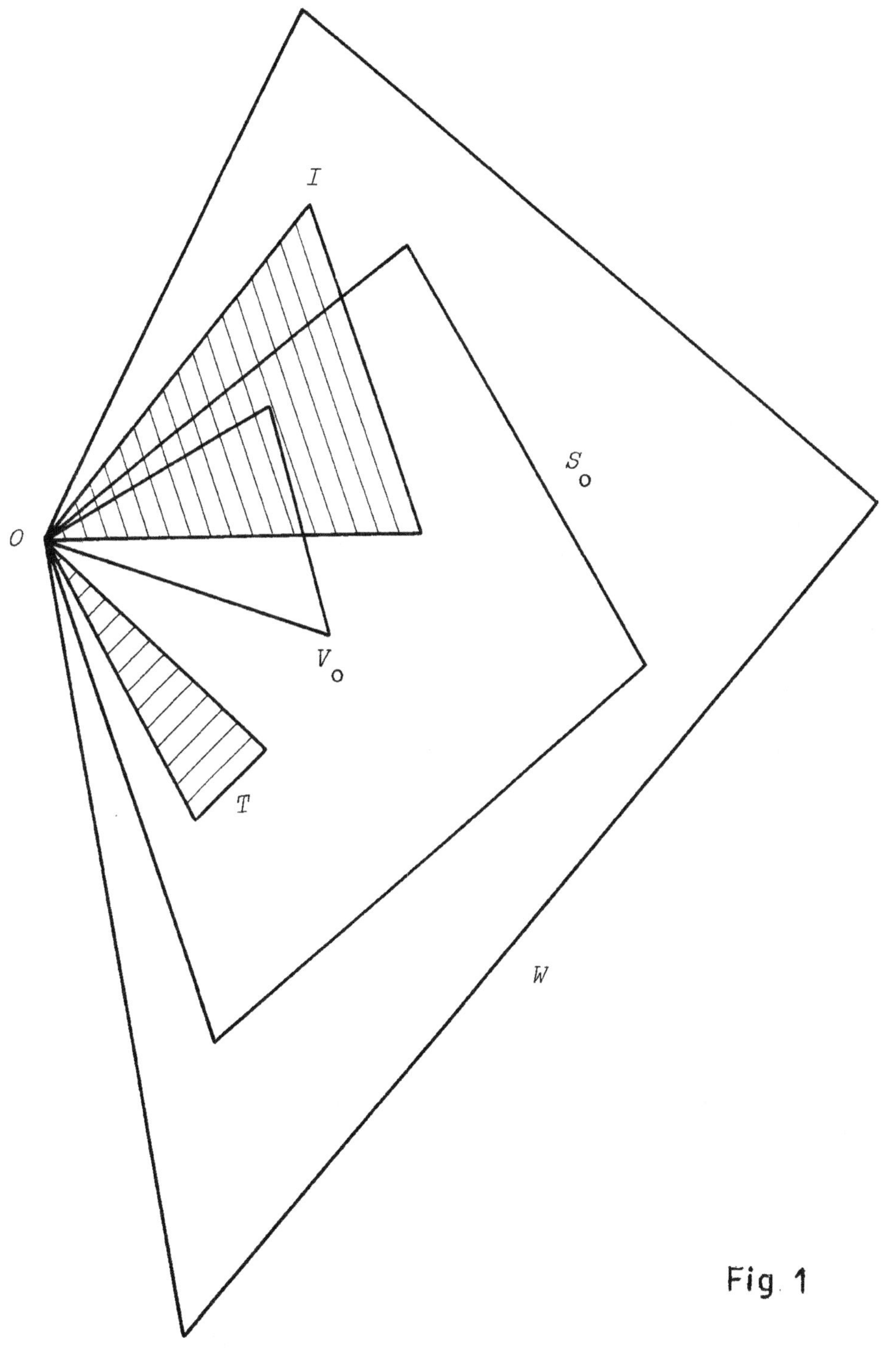

Fig. 1

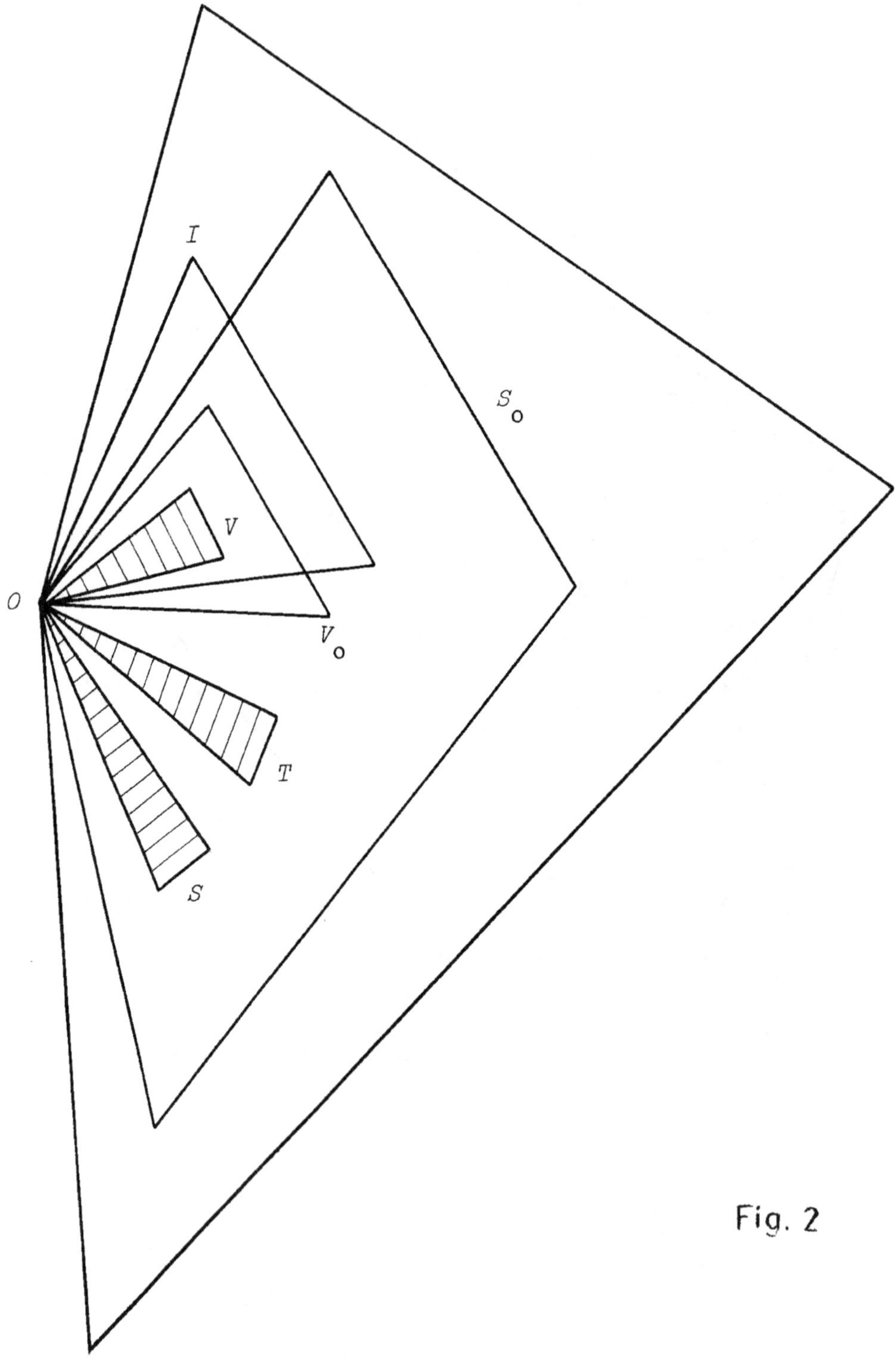

Fig. 2

a) compatible ideals I and vector subspaces T, as well as of

b) vector subspaces S_1 satisfying (6) and (7).

These two problems will be solved in Theorem 2, §5, respectively Corollary 2, §6.

§4. LOCALLY VANISHING IDEALS OF SEQUENCES OF SMOOTH FUNCTIONS

A first specialization of the ideals I in Theorem 1, §3 is given here, under the form of locally vanishing ideals.

For $p \in \bar{N}^n$ denote by W_p the set of all sequences of smooth functions $w \in W$ satisfying the local vanishing property

(12)
$$\begin{aligned} &\forall\ x \in R^n,\ q \in N^n,\ q \leq p : \\ &\exists\ \mu \in N : \\ &\forall\ \nu \in N,\ \nu \geq \mu : \\ &\quad D^q w(\nu)(x) = 0 \end{aligned}$$

or, formulated in a simpler way

(12') $\quad D^q w(\nu)(x) = 0$ for each $x \in R^n$, $q \in N^n$, $q \leq p$, if ν is big enough

Obviously W_p, with $p \in \bar{N}^n$, are ideals in W and $W_p \subset W^p$, $\forall\ p \in \bar{N}^n$ (see chap. 1 §10).

An ideal I in W is called locally vanishing, only if

(13) $\quad I \subset W_0$

Given a singularity generator Γ on R^n, a class of associated locally vanishing ideals is constructed now. For $G \subset F_\Gamma$ and $p \in \bar{N}^n$, denote by $I_{G,p}$ the ideal in W generated by all sequences of smooth functions $w \in W$ satisfying

(14) $\quad \exists\ G \in G :$

(14.1)
$$\begin{aligned} &\forall\ q \in N^n,\ q \leq p : \\ &\exists\ \mu_1 \in N : \\ &\forall\ \nu \in N,\ \nu \geq \mu_1 : \\ &\quad D^q w(\nu) = 0 \text{ on } G \end{aligned}$$

(14.2)
$$\begin{aligned} &\forall\ x \in R^n \backslash G : \\ &\exists\ V \text{ neighbourhood of } x,\ \mu_2 \in N : \\ &\forall\ \nu \in N,\ \nu \geq \mu_2 : \\ &\quad w(\nu) = 0 \text{ on } V \end{aligned}$$

or, formulated simply:

(14') $\exists\, G \in \mathcal{G}$:

(14'.1) $D^q w(\nu) = 0$ on G , for $q \in N^n$, $q \leq p$ and ν big enough,

(14'.2) $w(\nu) = 0$ on a neighbourhood of each $x \in R^n \setminus G$, if ν is big enough

In case $\mathcal{G} = \{G\}$, the notation $\mathcal{I}_{\mathcal{G},p} = \mathcal{I}_{G,p}$ will be used.

Proposition 1

$\mathcal{I}_{\mathcal{G},p} \subset \mathcal{W}_p$, therefore $\mathcal{I}_{\mathcal{G},p}$ is a locally vanishing ideal.

Proof

It suffices to show that $w \in \mathcal{W}_p$ whenever $w \in \mathcal{W}$ satisfies (14). Assume $w \in \mathcal{W}$ satisfies (14) for a certain $G \in \mathcal{G}$ and take $x \in R^n$. If $x \in G$ then (14.1) will imply (12). In case $x \in R^n \setminus G$, (12) will be implied by (14.2) $\nabla\nabla\nabla$

Denote by $\mathcal{J}_{\mathcal{G},p}$ the set of all sequences of smooth function $w \in \mathcal{W}$ satisfying (14). Obviously, $\mathcal{I}_{\mathcal{G},p}$ is the set of all finite sums of elements in $\mathcal{J}_{\mathcal{G},p}$.
Two examples of elements in $\mathcal{J}_{\mathcal{G},p}$ and thus, in $\mathcal{I}_{\mathcal{G},p}$, are presented in Lemmas 1 and 2
Suppose $w \in \mathcal{W}$, $\gamma \in \Gamma$, $\alpha \in \mathcal{C}^\infty(R^{m_\gamma})$ and define $w_{\gamma,\alpha} \in \mathcal{W}$ by

$$w_{\gamma,\alpha}(\nu)(x) = \alpha((\nu+1)\gamma(x)) \cdot w(\nu)(x) \ , \quad \forall \ \nu \in N \ , \ x \in R^n \ .$$

Lemma 1

If $\alpha \in \mathcal{D}(R^{m_\gamma})$ and satisfies for a given $k \in \bar{N}$ the condition

$$D^r\alpha(0) = 0 \ , \quad \forall \ r \in N^{m_\gamma} \ , \ |\, r \,| \leq k$$

then $w_{\gamma,\alpha} \in \mathcal{J}_{\mathcal{G},p}$, $\forall\ \mathcal{G} \subset \mathcal{F}_\Gamma$, $\mathcal{G} \ni F_\gamma$, $p \in \bar{N}^n$, $|\, p \,| \leq k$.

Proof

It can be seen that $w_{\gamma,\alpha}$ and $F_\gamma \in \mathcal{G}$ satisfy (14) $\nabla\nabla\nabla$

Suppose now $\gamma \in \Gamma$, with $m_\gamma = 1$ and denote by δ_γ the Dirac δ distribution of the surface F_γ . Suppose $s_\gamma \in \mathcal{S}_0$ such that $\langle s_\gamma , \cdot \rangle = \delta_\gamma$. For $q \in N^n$ and $\alpha,\beta \in \mathcal{C}^\infty(R^1)$ define $s_{\gamma,q} \in \mathcal{W}$ by

$$s_{\gamma,q}(\nu)(x) = \alpha((\nu+1)\gamma(x)) \cdot \beta(\gamma(x)) \cdot D^q s_\gamma(\nu)(x) \ , \quad \forall \ \nu \in N \ , \ x \in R^n$$

Lemma 2

If $\alpha \in \mathcal{D}(R^1)$, $\alpha = 1$ in a neighbourhood of $0 \in R^1$ and β satisfies for a given $k \in \bar{N}$ the condition

$$D^r\beta(0) = 0 \ , \quad \forall \ r \in N \ , \quad r \leq k$$

then

1) $s_{\gamma,q} \in \mathcal{J}_{G,p} \cap \mathcal{S}_o \ , \quad \forall \ G \subset F_\Gamma \ , \quad G \ni F_\gamma \ , \quad p \in \bar{N}^n \ , \quad |p| \leq k \ ,$

2) $s_{\gamma,q} \in \mathcal{V}_o$ provided that $|q| < k$.

Proof

1) It can be seen that $s_{\gamma,q}$ and $F_\gamma \in \mathcal{G}$ satisfy (14), therefore $s_{\gamma,q} \in \mathcal{J}_{G,p}$. The relation $s_{\gamma,q} \in \mathcal{S}_o$ results easily.

2) It follows easily ∇∇∇

An important property of the sequences of smooth functions $s \in \mathcal{I}_{G,p} \cap \mathcal{S}_o$ is given in:

Proposition 2

Suppose $s \in \mathcal{I}_{G,p} \cap \mathcal{S}_o$ then

$\exists \ G_1 \ , \ \dots \ , \ G_h \in \mathcal{G}$:

$\text{supp} <s,\cdot> \subset \text{fr } G_1 \cup \dots \cup \text{fr } G_h$ *)

Therefore $\text{int supp} <s,\cdot> = \emptyset$. *)

Proof

Since $s \in \mathcal{I}_{G,p}$, there exist $w_1 , \dots , w_h \in \mathcal{J}_{G,p}$ and $G_1 , \dots , G_h \in \mathcal{G}$ such that

(15) $\quad s = w_1 + \dots + w_h$

and w_i , G_i , with $1 \leq i \leq h$, satisfy (14). Since $G_1 , \dots , G_h$ are closed, the relations $s \in \mathcal{S}_o$, (15) and (14.2) imply

(16) $\quad \text{supp} <s,\cdot> \subset G_1 \cup \dots \cup G_h$

Take now $1 \leq i \leq h$ and $x \in \text{int } G_i$ and denote

$$I = \{ 1 \leq j \leq h \mid x \in \text{int } G_j \} , \quad J = \{ 1 \leq j \leq h \mid x \in \text{fr } G_j \} ,$$
$$K = \{ 1 \leq j \leq h \mid x \notin G_j \}$$

Obviously

$$I \cap J = I \cap K = J \cap K = \emptyset , \quad I \cup J \cup K = \{ 1,\dots,h \}$$

For $j \in I$ take $V_j \subset G_j$, V_j neighbourhood of x . For $j \in K$ take $V_j \subset R^n \setminus G_j$, V_j the neighbourhood of x resulting from (14.2). Denote $V = \bigcap_{j \in I \cup K} V_j$. Then (14.1)

---

*) fr A and int A denote respectively the frontier and interior of a subset $A \subset R^n$

applied to V_j with $j \in I$ and (14.2) applied to V_j with $j \in K$ will result through (15) in

$$\exists\ \mu \in N : \quad \forall\ \nu \in N\ ,\ \nu \geq \mu\ ,\ x \in V : s(\nu)(y) = \sum_{j \in J} w_j(\nu)(y)$$

If $J = \emptyset$, the above relation implies $x \notin \text{supp} <s,\cdot>$. Assume $J \neq \emptyset$ and $j \in J$, then $x \in \text{fr}\ G_j$. Taking now into account (16), the proof is completed ∇∇∇

Corollary 1

Suppose $\mathcal{G} \subset \mathcal{F}_{\Gamma,loc}$ (see §2), $p \in \bar{N}^n$ and $s \in \mathcal{I}_{\mathcal{G},p} \cap \mathcal{S}_o$ then

$\exists\ \Delta_s \subset \Gamma$:

1) $(F_\gamma \mid \gamma \in \Delta_s)$ locally finite

2) $\text{supp} <s,\cdot> \subset \bigcup_{\gamma \in \Delta_s} \text{fr}\ F_\gamma$

Proof

For each $G_i \in \mathcal{G}$ in Proposition 2, there exists $\Delta_i \subset \Gamma$ such that $(F_\gamma \mid \gamma \in \Delta_i)$ locally finite and $G_i = \bigcup_{\gamma \in \Delta_i} F_\gamma$, therefore $\text{fr}\ G_i \subset \bigcup_{\gamma \in \Delta_i} \text{fr}\ F_\gamma$.

Choosing $\Delta = \Delta_1 \cup \ldots \cup \Delta_h$, the proof is completed ∇∇∇

§5. LOCAL CLASSES AND COMPATIBILITY

A specialization of the vector spaces $\mathcal{T}$ in Theorem 1, §3, is given now.

A vector subspace $\mathcal{T}$ in $\mathcal{S}_o$ is called a local class, only if

(17.1) $\quad \mathcal{T} \cap \mathcal{V}_o = \mathcal{O}$

(17.2) $\quad \forall\ t \in \mathcal{T}\ ,\ t \neq \mathcal{O}$:

$\exists\ x \in R^n$:

$\forall\ \mu \in N$:

$\exists\ \nu \in N\ ,\ \nu \geq \mu$:

$t(\nu)(x) \neq 0$

Proposition 3

A locally vanishing ideal $\mathcal{I}$ and a local class $\mathcal{T}$ are compatible, only if

$$\mathcal{I} \cap \mathcal{S}_o \subset \mathcal{V}_o \oplus \mathcal{T}$$

Proof

It suffices to show that $\mathcal{I}$ and $\mathcal{V}$ satisfy (4). Taking now into account (17.1) it remains to prove that $I \cap T = 0$. Assume $v \in I \cap T$, $v \notin 0$.

Then (17.2) results in

$$\exists\ x \in R^n : \forall\ \mu \in N : \exists\ \nu_\mu \in N ,\ \nu_\mu \geq \mu : \ v(\nu_\mu)(x) \neq 0$$

But (13) and (12) with $q = 0$ will imply that

$$\exists\ \mu' \in N : \forall\ \nu \in N ,\ \nu \geq \mu' : \ v(\nu)(x) = 0$$

The contradiction obtained ends the proof ∇∇∇

A basic result, implying the existence of compatible locally vanishing ideals and local classes is given in the following proposition whose proof uses the cardinal equivalence between R^1 and $C^0(R^n)$.

Proposition 4

For any vector subspace J in $W_0 \cap S_0$ there exist local classes T , such that $J \subset S_0 \oplus T$.

Proof

Assume $(e_i \mid i \in I)$ is a Hamel base in the vector space $E = J/(J \cap V_0)$. Then $e_i = s_i + (J \cap V_0)$, where $s_i \in J$. Assume $\psi \in C^\infty(R^n)$ such that $\psi(x) \neq 0$, $\forall\ x \in R^n$. One can assume the existence of an injective mapping $a : I \to (-1,1)$. Indeed

$$J \subset W \subset (N \to C^0(R^n))$$

and

$$\text{car } C^0(R^n) = \text{car } R^1 \quad *)$$

therefore

$$\text{car } E \leq \text{car } J \leq (\text{car } R^1)^{\text{car } N} = \text{car } R^1$$

Now, define $v_i \in S_0$, with $i \in I$, by the relation

(18) $$v_i(\nu)(x) = (a(i))^\nu \psi(x) , \quad \forall\ \nu \in N ,\ x \in R^n .$$

Denote by T the vector subspace generated in S_0 by $\{ s_i + v_i \mid i \in I \}$. We prove that T is the sought after local class.

First, the inclusion $J \subset V_0 + T$. Assume $s \in J$ then $s + J \cap V_0 = \sum_{i \in J} c_i e_i$, for certain $J \subset I$, J finite and $c_i \in C^1$. Hence $s - \sum_{i \in J} c_i s_i = v \in J \cap V_0$. Denoting $t = \sum_{i \in J} c_i(s_i + v_i)$ it follows that $t \in T$ and $s = v - \sum_{i \in J} c_i v_i + t \in V_0 + T$, since $v_i \in V_0$ with $i \in J$.

---

*) car X denotes the cardinal number of the set X

It only remains to prove that $\mathcal{T}$ is a local class. First, the relation (17.1). Assume $t \in V_0 \cap \mathcal{T}$. The relation $t \in \mathcal{T}$ implies

(19) $$t = \sum_{i \in J} c_i(s_i + v_i)$$

where $J \subset I$, J finite and $c_i \in C^1$. Now, $t \in V_0$ results in $\sum_{i \in J} c_i s_i \in V_0$, since $v_i \in V_0$, with $i \in J$. But $\sum_{i \in J} c_i s_i \in \mathcal{J}$, hence $\sum_{i \in J} c_i e_i = 0 \in E$, which gives $c_i = 0$, $\forall\, i \in J$. Then, (19) will imply $t \in \mathcal{O}$. We prove now that (17.2) holds for any $t \in \mathcal{T}$, $t \notin \mathcal{O}$ and $x \in R^n$. Indeed, assume that

(20) $$\exists\ \mu \in N : \forall\ \nu \in N\ ,\ \nu \geq \mu\ :\ t(\nu)(x) = 0$$

for t given in (19). Since J is finite and $s_i \in \mathcal{J} \subset W_0$, with $i \in J$, the relation (12) with $q = 0$, will imply

(21) $$\exists\ \mu' \in N : \forall\ \nu \in N\ ,\ \nu \geq \mu'\ ,\ i \in J\ :\ s_i(\nu)(x) = 0$$

The relations (20), (21) and (19) give

(22) $$\exists\ \mu'' \in N : \forall\ \nu \in N\ ,\ \nu \geq \mu''\ :\ \sum_{i \in J} c_i v_i(\nu)(0) = 0$$

Taking into account (18) and the fact that $\psi(x) \neq 0$, we obtain from (22) the relations

(23) $$\sum_{i \in J} c_i (a(i))^{\nu} = 0\ ,\quad \forall\ \nu \in N\ ,\ \nu \geq \mu''$$

Since a is injective, (23) implies $c_i = 0$, $\forall\, i \in J$, therefore $t \in \mathcal{O}$, according to (20). The contradiction obtained ends the proof ∇∇∇

Now, the answer to the first problem in Remark 2, §3.

Theorem 2

For any locally vanishing ideal $\mathcal{I}$ there exist compatible local classes $\mathcal{T}$.

Proof

Assume $\mathcal{I}$ is a locally vanishing ideal. Denote $\mathcal{J} = \mathcal{I} \cap S_0$ then $\mathcal{J}$ is a vector subspace in $W_0 \cap S_0$ according to (13). Now, Proposition 4 will imply the existence of a local class $\mathcal{T}$ such that $\mathcal{J} \subset V_0 \oplus \mathcal{T}$. Taking into account the relation $\mathcal{J} = \mathcal{I} \cap S_0$ and Proposition 3, the above inclusion is the necessary and sufficient condition for the compatibility of $\mathcal{I}$ and $\mathcal{T}$. ∇∇∇

§6. DIRAC ALGEBRAS

The solution of the second problem in Remark 2, §3, namely, the existence of vector subspaces S_1 in S_o, satisfying (6) and (7), is obtained in Corollary 2, below.

Proposition 5

Suppose V and T are vector subspaces in V_o, respectively in S_o and the conditions

1) $V_o \cap T = 0$

2) $U \cap (V_o \oplus T) \subset U \cap (V(p) \oplus T)$, $\forall \quad p \in \mathbb{N}^n$

are satisfied.

Then, there exist vector subspaces S_1 in S_o so that (6) and (7) hold.

Proof

Denote $U_1 = U \cap (V_o \oplus T)$ and assume U_2 vector subspace in U such that $U = U_1 \oplus U_2$. Then $U_2 \cap (V_o \oplus T) = 0$, therefore, there exist vector subspaces S_2 in S_o such that $V_o \oplus T \oplus U_2 \oplus S_2 = S_o$. One can take now $S_1 = U_2 \oplus S_2$ ∇∇∇

A local class T is called <u>Dirac class</u>, only if

(24) $\quad \forall \quad t \in T \quad : \quad \text{int supp} \langle t, \cdot \rangle = \emptyset$

Corollary 2

Suppose T is a Dirac class, then there exist vector subspaces S_1 in S_o, satisfying (6) and the following stronger version of (7):

(25) $\quad U \subset S_1$

Proof

Due to (17.1) and (24) it follows easily that $U \cap (V_o \oplus T) = 0$, therefore, the conditions in Proposition 5 are satisfied. One can choose now S_1 as in the proof of the mentioned proposition ∇∇∇

The problem of finding Dirac classes T is solved in Proposition 6, below.

A locally vanishing ideal I is called <u>Dirac ideal</u>, only if

(26) $\quad \forall \quad s \in I \cap S_o \quad : \quad \text{int supp} \langle s, \cdot \rangle = \emptyset$

Theorem 3

For any Dirac ideal I there exist compatible Dirac classes T .

Proof

T constructed with $J = I \cap S_o$ according to the procedure in the proof of Proposition 4, will be a Dirac class. Moreover, due to Proposition 3, I and T will be compatible (see the proof of Theorem 2, § 5) ∇∇∇

The last problem, namely to secure Dirac ideals is solved in

Proposition 6

$I_{G,p}$ is a Dirac ideal, for any $G \subset F_\Gamma$ and $p \in \bar{N}^n$.

Proof

It follows from Proposition 2, § 4 ∇∇∇

Now, one can sum up the previous results and obtain the final answer on the existence of P-regularizations (V,S') for any admissible property P.

Theorem 4

For any Dirac ideal I (see Proposition 6) there exists a compatible Dirac class T. Further, there exist vector subspaces S_1 in S_o , satisfying the conditions:

(27) $$V_o \oplus T \oplus S_1 = S_o$$

(28) $$U \subset S_1$$

Choosing any vector subspace V in $I \cap V_o$, one obtains $(V,T \oplus S_1) \in R(P)$ for all admissible properties P.

Proof

Assume given a Dirac ideal I , for instance, according to the method in Proposition 6 Then Theorem 3 grants the existence of a compatible Dirac class T . Now, according to Corollary 2 , one can obtain a vector subspace S_1 in S_o satisfying (27) and (28). Taking into account Theorem 1, §3, the proof is completed ∇∇∇

The algebras used in the applications presented in chapters 3, 4 and 5 can be defined now.

Suppose given a Dirac ideal I_1 and a compatible Dirac class T_1 . For any ideal I in W , $I \supset I_1$, compatible vector subspace T in S_o , $T \supset T_1$, vector subspace V

in $I \cap V_o$ and vector subspace S_1 in S_o satisfying (6) and (7), the algebras (see (24) in chap. 1):

$$A^Q(V, T \oplus S_1, p), \quad p \in \bar{N}^n$$

where Q is a given admissible property, will be called Dirac algebras.

Remark C

The basic method whithin the present work, in constructing regularizations - and therefore, algebras containing the distributions - has been given in Theorem 1, §3. That method rests upon the notion of compatibility between an ideal I in W and a vector subspace T in S_o.

It is worthwhile mentioning the bottleneck feature the notion of compatibility exhibits Namely, given the ideal I, the compatible vector subspace T has to be small enough in order to satisfy (4) but in the same time, big enough in order to satisfy (5).

In that respect, Theorems 2 and 3 are nontrivial. Both of them are based on Proposition 4, which rests upon the cardinal equivalence between R^1 and $C^o(R^n)$, an essential characteristic of the set of real numbers.

Alternative ways, not depending on Dirac ideals and classes, but still within the framework of Theorem 1 in §3 of constructing regularizations will be given in §§8 and 9.

§7. MAXIMALITY

Taking into account §10 in chap. 1 as well as §3 above, it follows that there exists an applicative interest in constructing algebras containing the distributions, based on large, possibly maximal compatible ideals I and vector subspaces T.

Denote by C the set of all pairs (I,T) of compatible ideals I in W and vector subspaces T in S_o satisfying for a certain vector subspace S_1 in S_o the conditions (see Theorem 1, §3):

(29) $$V_o \oplus T \oplus S_1 = S_o ,$$

(30) $$U \subset V(p) \oplus T \oplus S_1 , \quad \forall \; p \in \bar{N}^n$$

where $V = I \cap V_o$.

Define a partial order $\leq$ on C by

$$(I_1, T_1) \leq (I_2, T_2) \Leftrightarrow I_1 \subset I_2 \text{ and } T_1 \subset T_2$$

Lemma 3

Each chain in $(C,\leq)$ has an upper bound.

Proof

Assume $((I_\lambda , T_\lambda) \mid \lambda \in \Lambda)$ is a chain in $(C,\leq)$ and denote $I = \bigcup_{\lambda\in\Lambda} I_\lambda$ and $T = \bigcup_{\lambda\in\Lambda} T_\lambda$, then obviously I is an ideal W, T is a vector subspace in S_0 and they are compatible. It only remains to show that I and T satisfy (29) and (30).

Denote

$$U_0 = U \cap \bigcap_{p\in\bar{N}^n} (V(p) \oplus T)$$

and assume U_1 is a vector subspace in U , such that $U = U_0 \oplus U_1$. If

(31) $$(V_0 \oplus T) \cap U_1 = O$$

then, one can choose a vector subspace S_2 in S_0 , such that

$$V_0 \oplus T \oplus U_1 \oplus S_2 = S_0$$

Denote $S_1 = U_1 \oplus S_2$, then (29) and (30) hold obviously. Now, if (31) is false, then there exists $\lambda \in \Lambda$ such that

(32) $$(V_0 \oplus T_\lambda) \cap U_1 \neq 0$$

But, $(I_\lambda , T_\lambda) \in C$, therefore there exists a vector subspace $S_{1\lambda}$ in S_0 such that

(33) $$V_0 \oplus T_\lambda \oplus S_{1\lambda} = S_0$$

(34) $$U \subset V_\lambda(p) \oplus T_\lambda \oplus S_{1\lambda} , \quad \forall \ p \in \bar{N}^n$$

where $V_\lambda = I_\lambda \cap V_0$. However, the relations (33) and (34) contradict (32). Indeed, denote

(35) $$U_{0\lambda} = U \cap \bigcap_{p\in\bar{N}^n} (V_\lambda(p) \oplus T_\lambda)$$

then, obviously $U_{0\lambda} \subset U_0$, therefore there exists a vector subspace $U_{1\lambda}$ in U , such that $U = U_{0\lambda} \oplus U_{1\lambda}$ and $U_{1\lambda} \supset U_1$. Then, (32) implies

$$(V_0 \oplus T_\lambda) \cap U_{1\lambda} \neq 0$$

Assume $u_1 \in U_{1\lambda}$, $u_1 \notin 0$, $v \in V_0$ and $t \in T_\lambda$ such that

(36) $$u_1 = v + t$$

But, (34) implies

(37) $$u_1 = v_\lambda + t' + s_1$$

where $v_\lambda \in \bigcap_{p\in\bar{N}^n} V_\lambda(p)$, $t' \in T_\lambda$ and $s_1 \in S_{1\lambda}$. Now, the relations (36), (37) and (33) result in $v = v_\lambda$, $t = t'$ and $s_1 \in O$. Then (37) gives $u_1 = v_\lambda + t'$ which together with $u_1 \in U_{1\lambda} \subset U$ and (35) will imply $u_1 \in U_{0\lambda}$. Therefore $u_1 \in O$, since $u_1 \in U_{1\lambda}$. The contradiction obtained completes the proof ∇∇∇

It follows, due to Zorn's lemma, that the set

$$C_{max} = \{ (I,T) \in C \mid (I,T) \text{ maximal in } (C,\leq) \}$$

is not void. Moreover,

$$\forall \quad (I,T) \in C :$$
$$\exists \quad (\bar{I},\bar{T}) \in C_{max} :$$
$$I \subset \bar{I} \; , \; T \subset \bar{T}$$

§8. LOCAL ALGEBRAS

In this section, regularizations (V,S') will be constructed according to the procedure in Theorem 1, §3, for the biggest locally vanishing ideal $I = W_0$ and for $V \subset I \cap V_0$. The resulting algebras however, will not be used in the present work and their interest here is only due to the alternative proof they offer for the existence of regularizations.

Proposition 7

Suppose $I = W_0$, then there exists a compatible local class T and a vector subspace S_1 in S_0 satisfying (6) and (25).

Proof

The existence of a compatible local class T results from Propositions 3 and 4 in §5. The problem is the existence of a suitable vector subspace S_1 in S_0 . That will be obtained through Corollary 2 in §6.

In this respect, we recall the way T was obtained in the proof of Proposition 4.

Assume $J = W_0 \cap S_0$ and $(e_i \mid i \in I)$ is a Hamel base in the vector space $E = J/(J \cap V_0)$. Then $e_i = s_i + W_0 \cap V_0$, with $s_i \in W_0 \cap S_0$. Now, T is obtained as the vector subspace generated in S_0 by $\{s_i+v_i \mid i \in I\}$ where v_i are given in (18).

We shall prove that

$$U \cap (V_0 \oplus T) = 0 \qquad (38)$$

Indeed, assume $\psi \in C^{\infty}(R^n)$, $v \in V_0$ and $t \in T$ such that $u(\psi) = v + t$. Then, taking into account the definition of T and the fact that $v_i \in V_0$, one obtains

$$u(\psi) = w + \sum_{i \in J} c_i \, s_i \qquad (39)$$

where $w \in V_0$, $J \subset I, J$ finite and $c_i \in C^1$. But $s_i \in W_0$, Therefore, applying (12) for $q = 0$, one obtains from (39) the relation

$$\forall \ x \in R^n :$$
$$\exists \ \mu \in N :$$
$$\forall \ \nu \in N , \ \nu \geq \mu :$$
$$w(\nu)(x) = \psi(x)$$

which according to Lemma 4 below, gives $\psi = 0$ on R^n , thus ending the proof of (38).

Now, the relation (38) grants the existence of a vector subspace S_2 in S_o such that

$$V_o \oplus T \oplus U \oplus S_2 = S_o$$

Taking $S_1 = U \oplus S_2$, the proof is completed ∇∇∇

Lemma 4

Suppose $\psi \in C^o(R^n)$ and v is a sequence of continuous functions on R^n , weakly convergent to $0 \in D'(R^n)$ such that

$$\forall \ x \in R^n :$$
$$\exists \ \mu \in N :$$
$$\forall \ \nu \in N , \ \nu \geq \mu :$$
$$v(\nu)(x) = \psi(x)$$

then $\psi = 0$ on R^n .

Proof

Assume, it is false and $B \subset R^n$ is a nonvoid open subset such that $\psi(x) \neq 0$, $\forall \ x \in B$. But, according to Lemma 5 below, there exists $G \subset B$, G nonvoid, open and $\mu \in N$ such that

$$v(\nu)(x) = \psi(x) , \quad \forall \ x \in G , \ \nu \in N , \ \nu \geq \mu$$

It follows that for any $\chi \in D(R^n)$ with $\text{supp}\ \chi \subset G$, the relation holds

$$\int_{R^n} \psi(x)\chi(x)dx = \lim_{\nu\to\infty} \int_{R^n} v(\nu)(x)\chi(x)dx = \langle v,\chi \rangle = 0$$

which contradicts the fact that $\psi(x) \neq 0$, $\forall \ x \in G$ ∇∇∇

Lemma 5

Suppose E is a complete metric space and F is a topological space. Suppose given the continuous functions $f : E \to F$ and $f_\nu : E \to F$, with $\nu \in N$, such that

$\forall\ x \in E :$
$\exists\ \mu \in N :$
$\forall\ \nu \in N\ ,\ \nu \geq \mu :$
$f_\nu(x) = f(x)$

Then, for each nonvoid closed subset $H \subset E$, there exists a nonvoid relatively open subset $G \subset H$ and $\mu \in N$ such that

$$f_\nu(x) = f(x)\ ,\quad \forall\ x \in G\ ,\ \nu \in N\ ,\ \nu \geq \mu\ .$$

<u>Proof</u>

Given H and $\mu \in N$, denote

$$H_\mu = \{x \in H \mid f_\nu(x) = f(x)\ ,\ \forall\ \nu \in N\ ,\ \nu \geq \mu\}$$

The hypothesis implies obviously

$$\bigcup_{\mu \in N} H_\mu = H$$

Now, it is easy to notice that H_μ , with $\mu \in N$, are closed in E due to the continuity of f and f_ν . Since H is in itself a complete metric space, the Baire category argument implies the existence of $\mu_0 \in N$ such that the relative interior of H_{μ_0} is not void ▽▽▽

The alternative proof for the existence of regularizations, not based on Dirac ideals and classes is obtained in

<u>Theorem 5</u>

There exist local classes T compatible with W_0 as well as vector subspaces S_1 in S_0 satisfying

(40) $$V_0 \oplus T \oplus S_1 = S_0$$

(41) $$U \subset S_1$$

Choosing any vector subspace V in $W_0 \cap V_0$, one obtains a regularization $(V, T \oplus S_1)$.

<u>Proof</u>

It follows from Proposition 7 as well as Theorem 1 in §3 ▽▽▽

Suppose given an ideal I_* in W and $I_* \subset W_0$.

For any ideal I in W , $I \supset I_*$, compatible vector subspace T in S_0 , vector subspace V in $I \cap V_0$ and vector subspace S_1 in S_0 satisfying (6) and (7), the algebras

$$A^Q(V,T\oplus S_1\ ,\ p)\ ,\quad p \in \bar{N}^n$$

where Q is an admissible property, will be called local algebras. Obviously, they contain as particular cases the Dirac algebras.

Remark M

Theorem 6 in chap. 1, §10, the inclusion $W_o \subset W^o$ as well as §7 of the present chapter rise the question:

Are the local algebras obtained for $I = I_* = W_o$ maximal either in the sense that $V = W_o \cap V_o$ is maximal according to chap. 1, §10, or $I = W_o$ is maximal according to §7?

The algebras constructed in the next section give a negative answer.

§9. FILTER ALGEBRAS

Given a filter base $\mathcal{B}$ on R^n, denote by $W_{\mathcal{B}}$ the set of all sequences of smooth functions $w \in W$ which satisfy the condition

$$(42)\qquad \begin{aligned} &\exists\ B \in \mathcal{B} :\\ &\forall\ x \in B :\\ &\exists\ \mu \in N :\\ &\forall\ \nu \in N\ ,\ \nu \geq \mu :\\ &\quad w(\nu)(x) = 0 \end{aligned}$$

or, under a simpler form

$$(42')\qquad \begin{aligned} &\exists\ B \in \mathcal{B} :\\ &\quad w(\nu)(x) = 0\ ,\quad \forall\ x \in B\ ,\ \nu \in N\ ,\ \nu \text{ big enough} \end{aligned}$$

Obviously, $W_{\mathcal{B}}$ is an ideal in W.

If $\mathcal{B}_1$ and $\mathcal{B}_2$ are filter bases on R^n and $\mathcal{B}_2$ generates a larger filter than $\mathcal{B}_1$, then obviously $W_{\mathcal{B}_1} \subset W_{\mathcal{B}_2}$.

A filter base $\mathcal{B}$ on R^n is called strongly dense, only if $R^n \setminus B$ is nowhere dense in R^n for each $B \in \mathcal{B}$.

The following filters on R^n

$$\mathcal{F}_v = \{ R^n \}$$

$$\mathcal{F}_f = \{ F \subset R^n \mid R^n \setminus F \text{ finite } \}$$

$$\mathcal{F}_{\ell f} = \{ F \subset R^n \mid R^n \setminus F \text{ locally finite } \}$$

$$\mathcal{F}_{nd} = \{ F \subset R^n \mid R^n \setminus F \text{ nowhere dense } \}$$

are examples of strongly dense filter bases on R^n . Obviously, $F_v \subset F_f \subset F_{\ell f} \subset F_{nd}$. Moreover, if B is a strongly dense filter base on R^n , then $B \subset F_{nd}$.

Due to the relation

$$W_o = W_{F_v}$$

the algebras constructed in this section will contain as particular cases the local al gebras defined in §8.

And now the important property of the ideals W_B .

Proposition 8

> Suppose B is a strongly dense filter base on R^n . Then, there exist vector subspaces T in S_o compatible with the ideal $I = W_B$. Further, there exist vector subspaces S_1 in S_o satisfying (6) and (25).

Proof

We shall adapt the proof of Propositions 4 and 7.

Assume $(e_i \mid i \in I)$ is a Hamel base in the vector space $E = (I \cap S_o) / (I \cap V_o)$. Then $e_i = s_i + I \cap V_o$, where $s_i \in I \cap S_o$. Assume $\psi \in C^\infty(R^n)$ such that $\psi(x) \neq 0$, $\forall x \in R^n$. Finally, assume $a : I \to (-1,1)$ injective. Define now $v_i \in V_o$, with $i \in I$, by the relation

(43) $$v_i(\nu)(x) = (a(i))^\nu \psi(x) , \quad \forall \nu \in N , \quad x \in R^n .$$

Denote by T the vector subspace generated in S_o by $\{s_i + v_i \mid i \in I\}$. We shall prove that I and T are compatible.

The relations $I \cap S_o \subset V_o + T$ and $V_o \cap T = 0$ result easily, as can be seen in the proof of Proposition 4, in §5.

It only remains to prove that

(44) $$I \cap T = 0$$

Assume $t \in I \cap T$, then $t \in T$ implies

(45) $$t = \sum_{i \in J} c_i (s_i + v_i)$$

with $J \subset I$, J finite and $c_i \in C^1$. Now $t \in I = W_B$ implies

(46) $$\begin{aligned} &\exists \; B \in B : \\ &\forall \; x \in B : \\ &\exists \; \mu \in N : \\ &\forall \; \nu \in N , \quad \nu \geq \mu : \\ &\quad t(\nu)(x) = 0 \end{aligned}$$

In the same time, $s_i \in I = W_B$, with $i \in J$, and the finiteness of J imply

$$(47)\quad \begin{array}{l} \exists\ B' \in \mathcal{B} : \\ \forall\ x \in B' : \\ \exists\ \mu' \in N : \\ \forall\ \nu \in N,\ \nu \geq \mu',\ i \in J : \\ \quad s_i(\nu)(x) = 0 \end{array}$$

The relations (46), (47) and (45) result in

$$(48)\quad \begin{array}{l} \exists\ B'' \in \mathcal{B} : \\ \forall\ x \in B'' : \\ \exists\ \mu'' \in N : \\ \forall\ \nu \in N,\ \nu \geq \mu'' : \end{array}$$

$$(48.1)\qquad \sum_{i \in J} c_i\ v_i(\nu)(x) = 0$$

Due to (43), the relation (48.1) can be written as

$$(48.2)\qquad \sum_{i \in J} c_i\ (a(i))^{\nu} = 0$$

since $\psi(x) \neq 0$, $\forall\ x \in R^n$, and $B'' \in \mathcal{B}$ implies $B'' \neq \emptyset$. Further, a is injective, therefore (48.2) results in $c_i = 0$, $\forall\ i \in J$. Thus (45) will give $t \in \mathcal{O}$, ending the proof of (44) and establishing that $\mathcal{I}$ and $\mathcal{T}$ are compatible.

Now, we prove the second part of Proposition 8, namely the existence of suitable vector subspaces $\mathcal{S}_1$ in $\mathcal{S}_o$.

First, we prove the relation

$$(49)\qquad \mathcal{U} \cap (\mathcal{V}_o \oplus \mathcal{T}) = \mathcal{O}$$

Assume indeed $\psi \in \mathcal{C}^{\infty}(R^n)$, $v \in \mathcal{V}$ and $t \in \mathcal{T}$ given by (45) and such that $u(\psi) = v + t$ Then

$$(50)\qquad u(\psi) = w + \sum_{i \in J} c_i\ s_i$$

with $w \in \mathcal{V}_o$.

Now, (50) and (47) will give for a certain $B' \in \mathcal{B}$ the relation

$$(51)\quad \begin{array}{l} \forall\ x \in B',\ \nu \in N,\ \nu \geq \mu' : \\ \quad \psi(x) = w(\nu)(x) \end{array}$$

with μ' possibly depending on $x \in B'$. But $R^n \setminus B'$ is nowhere dense in R^n, since $B' \in \mathcal{B}$.

Therefore

$$(52)\quad \begin{array}{l} \forall\ G \subset R^n,\ G \neq \emptyset,\ \text{open} \\ \exists\ G' \subset G,\ G' \neq 0,\ \text{open} : \\ \quad G' \subset B' \end{array}$$

Now, (51), (52) and Lemma 5 in §8 imply that $\psi = 0$ on B', since $w \in V_o$ in (51). One can conclude therefore that $\psi = 0$ on R^n and the proof of (49) is completed.

The relation (49) implies the existence of a vector subspace S_2 in S_o such that

$$V_o \oplus T \oplus U \oplus S_2 = S_o$$

Taking $S_1 = U \oplus S_2$, the proof is completed ∇∇∇

Theorem 6

Given a strongly dense filter base B on R^n, there exist vector subspaces T in S_o compatible with W_B as well as vector subspaces S_1 in S_o satisfying

(53) $$V_o \oplus T \oplus S_1 = S_o$$

(54) $$U \subset S_1$$

Choosing any vector subspace V in $W_B \cap V_o$, one obtains a regularization $(V, T \oplus S_1)$.

Proof

It follows from Proposition 8 and Theorem 1 in §3 ∇∇∇

Suppose given a strongly dense filter base B on R^n and an ideal I_* in W such that $I_* \subset W_B$.

For any ideal I in W, $I \supset I_*$, compatible vector subspace T in S_o, vector subspace V in $I \cap V_o$ and vector subspace S_1 in S_o satisfying (6) and (7), the algebras

$$A^Q(V, T \oplus S_1, p), \quad p \in \bar{N}^n,$$

where Q is an admissible property, will be called filter algebras and they are within the present work the most general instances of algebras given by a specific construction.

The question in Remark M, §8, reformulated for the case of filter algebras obtained from $I = I_* = W_{F_{nd}}$, remains open.

§10. REGULAR ALGEBRAS

It is worthwhile noticing that the Dirac ideals $I_{G,p}$ (see §4 and Proposition 6 in §6), the locally vanishing ideals W_p (see (12)), the ideals W_B (see (42)) as well as the ideals I_δ and I^δ used in chapters 5, 6 and 7, are all subsequence invariant (chap 1, §6).

It will be shown in the present section (see Proposition 10) that starting with arbitrary subsequence invariant ideals I in W, one can under rather general conditions construct algebras containing the distributions.

An ideal I in W is called regular, only if

(55) $$U \cap (V_0 + I) = 0$$

and

(56)
$$\begin{aligned} &\forall \quad v \in I \cap V_0 : \\ &\exists \quad \mu \in N : \\ &\forall \quad \nu \in N, \quad \nu \geq \mu : \\ &\exists \quad x \in R^n : \\ &\quad v(\nu)(x) = 0 \end{aligned}$$

or, shortly,

(56') if $v \in I \cap V_0$ then $v(\nu)$ does not vanish in R^n for at most a finite number of $\nu \in N$

Proposition 9

Suppose I is a regular ideal. Then, there exist compatible vector subspaces T in S_0 and vector subspaces S_1 in S_0 satisfying (6) and (25).

Proof

We shall once more use the method of proof in Propositions 4, 7 and 8.

Assume $(e_i \mid i \in I)$ is a Hamel base in the vector space $E = (I \cap S_0) / (I \cap V_0)$. Then $e_i = s_i + I \cap V_0$ with $s_i \in I \cap S_0$. Assume $\psi \in C^\infty(R^n)$ such that $\psi(x) \neq 0$, $\forall\, x \in R^n$. Finally, assume $a : I \rightarrow (-1,1)$ injective and define $v_i \in V_0$, with $i \in I$, by

(57) $$v_i(\nu)(x) = (a(i))^\nu \cdot \psi(x), \quad \forall \quad \nu \in N, \quad x \in R^n$$

Denote by T the vector subspace in S_0 generated by $\{ s_i + v_i \mid i \in I \}$. We shall prove that I and T are compatible (see (4), (5)). First the relation

(58) $$V_0 \cap T = 0$$

Assume $v \in V_0 \cap T$, then $v \in T$ implies

(59) $$v = \sum_{i \in J} c_i(s_i + v_i)$$

where $J \subset I$, J finite and $c_i \in C^1$. But (59) gives

(60) $$\sum_{i\in J} c_i s_i = v - \sum_{i\in J} c_i v_i \in I \cap V_o$$

since $s_i \in I$ and $v, v_i \in V_o$. Now (60) results in $\sum_{i\in J} c_i e_i = 0 \in E$, hence $c_i = 0$, $\forall\ i \in J$. Then (59) will give $v \in O$, ending the proof of (58).

Now, the relation

(61) $$I \cap S_o \subset V_o \oplus T$$

Assume $s \in I \cap S_o$, then $s + I \cap V_o \in E$, hence

(62) $$s + I \cap V_o = \sum_{i\in J} c_i e_i$$

for certain $J \subset I$, J finite and $c_i \in C^1$. But (62) can be written as

$$s - \sum_{i\in J} c_i s_i = v \in I \cap V_o$$

therefore

$$s = \sum_{i\in J} c_i (s_i + v_i) + v - \sum_{i\in J} c_i v_i \in T \oplus V_o$$

since $v, v_i \in V_o$ and (60) is proved.

In order to prove that I and T are compatible, it remains to show that

(63) $$I \cap T = O$$

Assume $t \in I \cap T$, then $t \in T$ implies

(64) $$t = \sum_{i\in J} c_i (s_i + v_i)$$

with $J \subset I$, J finite and $c_i \in C^1$. Hence

(65) $$v = \sum_{i\in J} c_i v_i = t - \sum_{i\in J} c_i s_i \in I \cap V_o$$

since $v_i \in V_o$ and $t, s_i \in I$. But (56) applied to $v \in I \cap V_o$ gives

$$\begin{aligned} &\exists\ \mu \in N : \\ &\forall\ \nu \in N ,\ \nu \geq \mu : \\ &\exists\ x \in R^n : \\ &\quad v(\nu)(x) = 0 \end{aligned}$$

which together with (65) results in

$$\begin{aligned} &\forall\ \nu \in N ,\ \nu \geq \mu :\ \exists\ x \in R^n : \\ &\quad \sum_{i\in J} c_i v_i(\nu)(x) = 0 \end{aligned}$$

Now (57) and the fact that $\psi(x) \neq 0$, $\forall\ x \in R^n$, will imply

(66) $$\sum_{i\in J} c_i (a(i))^{\nu} = 0 , \quad \forall\ \nu \in N ,\ \nu \geq \mu$$

The well known property of the Vandermonde determinants applied to (66) gives $c_i = 0$, $\forall \ i \in J$, hence $t \in \mathcal{O}$ due to (64) and the proof of (63) is completed.

Now, we prove the existence of vector subspaces S_1 in S_0 satisfying (6) and (25). First, we prove

(67) $$U \cap (V_0 \oplus T) = \mathcal{O}$$

Obviously $T \subset V_0 + I$, hence (67) follows from (55). Now, the existence of the required S_1 results easily from (67) ∇∇∇

Theorem 7

Given a regular ideal I in W, there exist vector subspaces T in S_0 compatible with I as well as vector subspaces S_1 in S_0 satisfying

(68) $$V_0 \oplus T \oplus S_1 = S_0$$

(69) $$U \subset S_1$$

Choosing any vector subspace V in $I \cap V_0$, one obtains a regularization $(V, T \oplus S_1)$.

Proof

It results from Proposition 9 and Theorem 1 in §3 ∇∇∇

The existence of regular ideals is granted by:

Proposition 10

A subsequence invariant ideal I in W is proper only if it satisfies (56). Therefore, a subsequence invariant, proper ideal I in W which satisfies (55) is regular.

Proof

It suffices to show that (56) holds whenever I is proper. Assume it is false and $v \in I \cap V_0$ such that

$$\forall \ \mu \in N : \exists \ \nu_\mu \in N, \ \nu_\mu \geq \mu : \ v(\nu_\mu) \neq 0 \ \text{on} \ R^n$$

Define $w \in W$ by $w(\mu) = v(\nu_\mu)$, $\forall \ \mu \in N$. Then $w \in I$, since I is subsequence invariant. But obviously $1/w \in W$, therefore

$$u(1) = w \cdot (1/w) \in I \cdot W \subset I$$

contradicting the fact that $I \subsetneq W$ ∇∇∇

It follows that the ideals mentioned at the beginning of this section are regular (in the case of the ideals W_B , the additional condition that B is a strongly dense filter base on R^n is needed).

One can easily notice that the set of regular ideals is chain complete, therefore, due to Zorn's lemma, there exist maximal regular ideals containing any given regular ideal.

Suppose given a regular ideal I_1 and an ideal I_* in W such that $I_* \subset I_1$.

For any ideal I in W , $I \supset I_*$, compatible vector subspace T in S_0 , vector subspace V in $I \cap V$ and vector subspace S_1 in S_0 satisfying (6) and (7), the algebras

$$A^Q(V,T \oplus S_1 ,p) \ , \quad p \in \bar{N}^n$$

where Q is an admissible property, will be called regular algebras.

The regular algebras will find an important application in chapter 3, §4, where a general solution scheme is established for a wide class of nonlinear partial differential equations.

Remark 3

The condition (55) in the definition of a regular ideal is needed in order to secure the condition (69) in Theorem 7 (see (7) in Theorem 1, §3 and (67) in the proof of Proposition 9, as well as (20.2) in chap. 1, §7).

However, due to Proposition 5 in §6, one can replace (55) by the weaker condition

$$(70) \qquad U \cap (V_0 + I) \subset U \cap (V(p) \oplus T) \ , \quad \forall \ p \in \bar{N}^n$$

where $V = I \cap V_0$ and T was constructed in the proof of Proposition 9, since in that case the relation holds

$$V_0 \oplus T = V_0 + I \cap S_0$$

Chapter 3

SOLUTIONS OF NONLINEAR PARTIAL DIFFERENTIAL EQUATIONS APPLICATION TO NONLINEAR SHOCK WAVES

§1. INTRODUCTION

It will be proved in §2 of this chapter that, the piece wise smooth weak solutions of nonlinear partial differential equations with polynomial nonlinearities and smooth coefficients, satisfy these equations in the usual algebraic sense, with the multiplication and derivatives defined in the Dirac algebras containing $\mathcal{D}'(R^n)$ introduced in chapter 2.

An application to the shock wave solutions of nonlinear hyperbolic partial differential equations will be given in §3.

When dealing with partial differential equations, one has to consider various nonvoid open subsets Ω in R^n and restrict the functions and distributions to such subsets. It is obvious that the construction of the algebras containing the distributions, carried out in chapters 1 and 2, remains valid for any $\Omega \subset R^n$, $\Omega \neq \emptyset$, open.

§2. POLYNOMIAL NONLINEAR PARTIAL DIFFERENTIAL OPERATORS AND SOLUTIONS

Given $\Omega \subset R^n$, $\Omega \neq \emptyset$, open, a partial differential operator $T(D)$ is called polynomial nonlinear on Ω , only if

$$(1) \qquad T(D)u(x) = \sum_{1\leq i\leq h} L_i(D)\, T_i\, u(x) \ , \qquad \forall\ u \in C^\infty(\Omega)\ ,\ x \in \Omega\ ,$$

where $L_i(D)$ are linear partial differential operators with smooth coefficients, while T_i are polynomials of the form

$$(2) \qquad T_i\, u(x) = \sum_{1\leq j\leq k_i} c_{ij}\,(x)(u(x))^j \ , \qquad \forall\ u \in C^\infty(\Omega)\ ,\ x \in \Omega\ ,$$

with c_{ij} smooth.

The polynomial nonlinear partial differential operator $T(D)$ is called homogeneous, only if $u(x) = 0$, for $x \in \Omega$, implies $T(D)u(x) = 0$, for $x \in \Omega$.

Obviously, the polynomial nonlinear partial differential operators are particular cases of the operators in chap. 1, §9.

The nonlinear hyperbolic operators studied in §3 are examples of homogeneous polynomial nonlinear partial differential operators. The same is the case of several types of nonlinear wave operators studied in recent literature, [5], [8-11], [91], [121], [122], as well as other nonlinear partial differential operators, [80].

A function $u : \Omega \to C^1$ is called a piece wise smooth weak solution of the equation

(3) $\quad T(D)u(x) = 0 , \quad x \in \Omega$

only if the following conditions (4), (5), (6) and (7) are satisfied:

There exists a set Δ of mappings $\gamma : \Omega \to R^{m_\gamma}$, with $\gamma \in C^\infty$, $m_\gamma \in N$, such that the set (see chap. 2, §2) $F_\Delta = \{ x \in R^n \mid \exists \gamma \in \Delta : \gamma(x) = 0 \in R^{m_\gamma} \}$ is closed, has zero Lebesque measure in R^n and

(4) $\quad u \in C^\infty(\Omega \setminus F_\Delta)$

If $k = \max \{ k_i \mid 1 \le i \le h \}$ (see (2)) then

(5) $\quad u^k$ is locally integrable on Ω

The weak solution property holds

(6) $$\int_\Omega \Big(\sum_{1 \le i \le h} T_i \, u(x) L_i^*(D)\psi(x)\Big)dx = 0 , \quad \forall \ \psi \in \mathcal{D}(\Omega) ,$$

where $L_i^*(D)$ is the formal adjoint of $L_i(D)$.

For each $\gamma \in \Delta$ there exists a bounded and balanced neighbourhood B_γ of $0 \in R^m$, such that

(7) $\quad \{ \gamma^{-1}(B_\gamma) \mid \gamma \in \Delta \}$ is locally finite in Ω .

The nonlinear hyperbolic partial differential equations studied in §3, are known, [133], [52], to possess piece wise smooth weak solutions in the above sense.

The main result of the present chapter is presented in

Theorem 1

Given a homogeneous polynomial nonlinear partial differential operator $T(D)$ defined on a nonvoid open subset $\Omega \subset R^n$ and a piece wise smooth weak solution $u : \Omega \to C^1$ of the equation

(8) $\quad T(D)u(x) = 0 , \quad x \in \Omega ,$

there exist regularizations $(\mathcal{V},\mathcal{S}')$ (see chap. 1, §7) such that for any admissible property Q , one obtains

1) $u \in A^Q(\mathcal{V},\mathcal{S}',p)$, $\forall \ p \in \bar{N}^n$,

2) in the case of derivative algebras, u satisfies the equation (8) in the usual algebraic sense, with the respective multiplication and derivatives within the algebras $A^Q(\mathcal{V},\mathcal{S}',p)$, $p \in \bar{N}^n$.

<u>Proof</u>

Assume given $\alpha_\gamma : R^{m_\gamma} \to [0,1]$, $\alpha_\gamma \in C^\infty$, for each $\gamma \in \Delta$, in such a way that

(9.1) $\alpha_\gamma = 0$ on a certain neighbourhood V_γ of $0 \in R^{m_\gamma}$,

(9.2) $\alpha_\gamma = 1$ on $R^{m_\gamma} \setminus B_\gamma$ (see (7))

For $\nu \in N$ and $x \in R^n$ define a <u>regularization</u> of the piece wise smooth weak solution u by

(10) $$s(\nu)(x) = \begin{cases} u(x) \cdot \prod_{\gamma \in \Delta} \alpha_\gamma((\nu+1)\gamma(x)) & \text{if } x \in \Omega \setminus F_\Delta \\ 0 & \text{if } x \in F_\Delta \end{cases}$$

We prove that

(11) $s \in W(\Omega)$

Assume $\nu \in N$ given. If $x \in \Omega \setminus F_\Delta$ then

(12) $\{ \gamma \in \Delta \mid (\nu+1)\gamma(x) \in B_\gamma \}$ finite.

Indeed, $(\nu+1)\gamma(x) \in B_\gamma$ only if $x \in \gamma^{-1} (\frac{1}{\nu+1} B_\gamma)$. Therefore, (12) will result from (7) and the fact that B_γ , with $\gamma \in \Delta$, are balanced. But (12) and (9.2) imply that the product $\prod_{\gamma \in \Delta} \alpha_\gamma((\nu+1)\gamma(x))$ in (10) contains only a finite number of factors $\neq 1$. Thus $s(\nu)$ is well defined on $\Omega \setminus F_\Delta$. Since $\Omega \setminus F_\Delta$ is open, one can take a compact neighbourhood V of x , $V \subset \Omega \setminus F_\Delta$. Then, as in (12), one obtaines

(13) $\{ \gamma \in \Delta \mid (\nu+1)\gamma(V) \cap B_\gamma \neq \emptyset \}$ finite.

Now, (13) and (10) imply that $s(\nu) \in C^\infty$ in x .

If $x \in F_\Delta$ then $\gamma(x) = 0$ for a certain $\gamma \in \Delta$. Take V a neighbourhood of x , such that $\gamma(V) \subset \frac{1}{\nu+1} V_\gamma$ (see (9.1)), then

(14) $s(\nu) = 0$ on V

according to (9.1) and (10). Therefore, $s(\nu) \in C^\infty$ in x and the proof of (11) is completed.

Define $v \in W(\Omega)$ by

(15) $v = T(D)s$

The sequence of smooth functions v is obviously measuring the <u>error</u> in (8) obtained by replacing u with its regularization s given in (10) and it plays the essential role in constructing the <u>ideals</u> $I^Q(V(p),S')$ of sequences of smooth functions needed in the construction of the algebras $A^Q(V,S',p)$ (see (24), chap. 1, §7).

We prove the relations

$$\begin{array}{ll} & \forall \ K \subset \Omega \setminus F_\Delta , \quad K \text{ compact :} \\ & \exists \ \mu \in N : \\ (16) & \forall \ \nu \in N , \quad \nu \geq \mu : \\ & \quad s(\nu) = u \ \text{ on } \ K \\ & \quad v(\nu) = 0 \ \text{ on } \ K \end{array}$$

Indeed, denote $\Delta_K = \{ \gamma \in \Delta \mid \gamma(K) \cap B_\gamma \neq \emptyset \}$, then Δ_K is finite due to (7). Denote $a = \inf \{ ||\gamma(x)||_\gamma \mid \gamma \in \Delta_K , x \in K \}$ *) then $a > 0$, since Δ_K is finite and $K \cap F_\Delta = \emptyset$, K compact. Obviously, there exists $\mu \in N$, such that

$$\sup_{x_\gamma \in B_\gamma} ||x_\gamma||_\gamma \leq \mu a , \quad \forall \ \gamma \in \Delta$$

Then $(\nu+1)\gamma(K) \subset R^{m_\gamma} \setminus B_\gamma$, $\forall \gamma \in \Delta , \nu \in N , \nu \geq \mu$. Now, (16) results easily from (10), (15) and (6).

An other relation needed, given in

$$\begin{array}{ll} (17) & \forall \ \nu \in N , \quad p \in \bar{N}^n : \\ & \quad D^p s(\nu) = D^p v(\nu) = 0 \ \text{ on } \ F_\Delta , \end{array}$$

results easily from (14) and (15), since $T(D)$ is homogeneous.

A last property of the regularization s , given in

$$(18) \qquad s \in \mathcal{S}_o(\Omega) \ \text{ and } \ \langle s,\cdot\rangle = u$$

follows obviously from (11), (16), (4) and (5), since in the last relation one can assume $k \geq 1$, otherwise $T(D)$ being trivial.

The preliminary results above lead to the following essential property of the error sequence v :

$$(19) \qquad v \in \mathcal{V}_o(\Omega) \ \text{ and } \ v \in \mathcal{I}_{F_\Delta , p} , \quad \forall \ p \in \bar{N}^n \quad \text{(see chap. 2, §3)}$$

Indeed, the relation $v \in \mathcal{I}_{F_\Delta ,p}$, $\forall \ p \in \bar{N}^n$ results from (16), (17).

It only remains to prove that $v \in \mathcal{V}_o(\Omega)$. Assume $\psi \in \mathcal{D}(\Omega)$ and $\nu \in N$, then (15) and (6) imply

$$\Big| \int_\Omega v(\nu)(x)\psi(x)dx \Big| = \Big| \int_\Omega \sum_{1\leq i\leq h} T_i \, s(\nu)(x) \, L_i^*(D)\psi(x)dx \Big| =$$

$$= \Big| \int_\Omega \sum_{1\leq i\leq h} (T_i s(\nu)(x) - T_i u(x)) \, L_i^*(D)\psi(x)dx \Big| \leq$$

$$\leq \sum_{1\leq i\leq h} \int_{\text{supp}\,\psi} | T_i \, s(\nu)(x) - T_i \, u(x) | \cdot | L_i^*(D)\psi(x) | \, dx$$

---

*) $|| \ ||_\gamma$ is a norm on R^{m_γ} , with $\gamma \in \Delta$, so that $\sup_{\gamma\in\Delta} \ \sup_{x_\gamma \in B_\gamma} ||x_\gamma||_\gamma < \infty$

Therefore, it suffices to prove

(20) $$\lim_{\nu\to\infty} \int_K | T_i s(\nu)(x) - T_i u(x) | \, dx = 0 , \quad \forall \; 1 \leq i \leq h , \quad K \subset \Omega , \quad K \text{ compact}$$

First, (2) and (10) imply for $x \in \Omega \setminus F_\Delta$ the relation

(21) $$T_i s(\nu)(x) - T_i u(x) = \sum_{1\leq j\leq k_i} c_{ij}(x)(u(x))^j \cdot \left(\prod_{\gamma\in\Delta} (\alpha_\gamma((\nu+1)\gamma(x)))^j - 1 \right) ,$$
$$\forall \; 1\leq i\leq h , \quad \nu \in N .$$

And, due to (9) one obtaines

(22) $$\left| \prod_{\gamma\in\Delta} (\alpha_\gamma((\nu+1)\gamma(x)))^j - 1 \right| \leq 1 , \quad \forall \; j \in N , \quad \nu \in N , \quad x \in \Omega ,$$

while taking into account also (7), it follows that

(23) $$\lim_{\nu\to\infty} \left(\prod_{\gamma\in\Delta} (\alpha_\gamma((\nu+1)\gamma(x)))^j - 1 \right) = 0 , \quad \forall \; j \in N , \quad x \in \Omega \setminus F_\Delta .$$

Now, (21), (22) and (23) together with (5) and the fact (see (4)) that the Lebesque measure of F_Δ in R^n is zero, will imply (20), completing the proof of (19).

The above relation (19) offers the possibility of constructing the ideals $I^Q(V(p),S')$ upon which the construction of the algebras $A^Q(V,S',p)$ is based.

Denote by I_v the ideal in $W(\Omega)$ generated by v , then

(24) I_v is a Dirac ideal (chap. 2, §5) and
$$I_v \subset I_{F_\Delta ,p} , \quad \forall \; p \in \bar{N}^n \quad \text{(chap. 2, §4)}$$

Indeed, according to (19), $v \in I_{F_\Delta ,p}$ therefore, $I_v \subset I_{F_\Delta ,p}$ and Proposition 6, chap. 2, §6, implies that I_v is a Dirac ideal.

Assume now given any Dirac ideal I , such that

(25) $$I_v \subset I$$

then, according to Theorem 4, chap. 2, §6, there exists a Dirac class T , compatible with I . Thus, there exists a vector subspace S_1 in $S_0(\Omega)$, satisfying the conditions

(26) $$V_0(\Omega) \oplus T \oplus S_1 = S_0(\Omega)$$

(27) $$U(\Omega) \subset S_1$$

If u is not smooth, then S_1 can be chosen so that

(28) $$s \in S_1$$

Indeed, in that case $s \notin V_0(\Omega) \oplus T \oplus U$ since $u = \langle s,\cdot\rangle$ is piece wise smooth and T is a Dirac class.

One can choose a vector subspace V in $I \cap V_0(\Omega)$ such that

(29) $$v \in V(p) , \quad \forall \; p \in \bar{N}^n$$

Indeed, (19) implies that

$$D^q v \in V_o(\Omega) \ , \quad \forall \ q \in N^n$$

while (16) and (17) and the fact that $\Omega \setminus F_\Delta$ is open, result in

$$D^q v \in I_{F_\Delta ,p} \ , \quad \forall \ q \in N^n \ , \quad p \in \bar{N}^n$$

Denote now

(30) $\qquad S' = T \oplus S_1$

then Theorem 4 in chap. 2, §6, will imply that (V,S') is a Q-regularization for any admissible property Q .

The relation

(31) $\qquad u \in A^Q(V,S',p) \ , \quad \forall \ p \in \bar{N}^n$

results easily from (18) and the fact that $D'(\Omega) \subset A^Q(V,S',p)$, with $p \in \bar{N}^n$.

It only remains to prove that u satisfies (8) in the usual algebraic sense, with the respective multiplication and derivatives in the algebras $A^Q(V,S',p)$.

Due to (28) and (30) one obtains

(32) $\qquad u = s + I^Q(V(p),S') \in A^Q(V,S',p) \ , \quad \forall \ p \in \bar{N}^n$

therefore, taking into account §9 and Theorems 2, 3, §8, chap. 1, as well as (15) and (29), one obtaines in the case of derivative algebras, the relations

(33) $\qquad T(D)u = T(D)s + I^Q(V(p),S') = v + I^Q(V(p),S') = 0 \in A^Q(V,S',p) \ , \quad \forall \ p \in \bar{N}^n$

The relations (32) and (33) end the proof of Theorem 1 $\nabla\nabla\nabla$

Remark 1

The regularizations (V,S') whose existence is stated in Theorem 1, are obtained in a rather simple, constructive way. Obviously, the algebras $A^Q(V,S',p)$ obtained are Dirac algebras.

§3. APPLICATION TO NONLINEAR SHOCK WAVES

Consider the nonlinear hyperbolic partial differential equation

(34) $\qquad u_t(x,t) + a(u(x,t)) \cdot u_x(x,t) = 0 \ , \quad x \in R^1 \ , \quad t > 0$

(35) $\qquad u(x,0) = u_o(x) \ , \quad x \in R^1$

where $a : R^1 \to R^1$ is a polynomial.

Obviously, the left part of (34) is a polynomial nonlinear partial differential operator on $\Omega = R^1 \times (0,\infty) \subset R^2$ and it is homogeneous.

Under rather general conditions, for smooth, [133], [52], or even piece wise smooth, [32], initial data u_0 , the equation (34), (35) possesses shock wave solutions $u : \Omega \to R^1$, with the properties:

There exists a finite set Δ of smooth curves $\gamma : \Omega \to R^1$, such that

(36) $\quad u \in C^\infty(\Omega \setminus F_\Delta)$

(37) $\quad u$ locally bounded on Ω

(38) $$\int_\Omega (u(x,t)\psi_t(x,t) + f(u(x,t)) \cdot \psi_x(x,t))dx\, dt = 0 , \quad \forall\ \psi \in D(\Omega)$$

where f is a primitive of a .

Obviously, such a solution u will be a piece wise smooth weak solution, in the sense of the definition in §2.

Therefore, Theorem 1 in §2 results in :

Theorem 2

If $u : \Omega \to R^1$ is a shock wave solution of (34), (35) which satisfies (36), (37) and (38), then, there exist regularizations (V,S') (see chap. 1, §7) such that for any admissible property Q , one obtains

1) $u \in A^Q(V,S',p)$, $\forall\ p \in \bar{N}^n$

2) in the case of derivative algebras, u satisfies (34) in the usual algebraic sense in each of the algebras $A^Q(V, S',p)$, $p \in \bar{N}^n$, with the respective multiplication and derivatives.

§4. GENERAL SOLUTION SCHEME FOR NONLINEAR PARTIAL DIFFERENTIAL EQUATIONS

With the help of the regular algebras introduced in chapter 2, §10, the general result on the piece wise smooth weak solutions of homogeneous polynomial nonlinear partial differential equations, established in Theorem 1, §2, can be seen as a particular case of a yet more general solution scheme for a fairly arbitrary class of nonlinear partial differential equations presented next, in Theorem 4.

Suppose given a nonlinear partial differential operator (see chap. 1, §9) of the general form

(39) $$T(D)u(x) = \sum_{1 \le i \le h} c_i(x) \prod_{1 \le j \le k_i} D^{p_{ij}}u(x) , \quad x \in \Omega ,$$

where $\Omega \subset R^n$ is nonvoid and open, and $c_i \in C^\infty(\Omega)$, $p_{ij} \in N^n$.

A distribution $S \in D'(\Omega)$ is called a (nontrivial) regular weak solution of the nonlinear partial differential equation

(40) $$T(D)u(x) = 0 , \quad x \in \Omega ,$$

only if there exists a weakly convergent sequence of smooth functions $s \in \mathcal{S}_o(\Omega)$ satisfying the following three conditions

(41) $S = \langle s,\cdot\rangle$ and there exists a nonvoid open subset $G \subset \Omega$ and a smooth function $u \in C^\infty(G)$, such that

(41.1) $S = u = s(\nu)$ on G , $\forall\ \nu \in N$,

further

(42) $v = T(D)s \in \mathcal{V}_o(\Omega)$

and finally, the ideal $\mathcal{I}_v$ generated in $\mathcal{W}(\Omega)$ by $D^p v$, with $p \in N^n$, satisfies the condition

(43) $\mathcal{I}_v \cap (\mathcal{V}_o + \mathcal{U} + \mathcal{R}) \subset \mathcal{V}_o$, where $\mathcal{R} = C^1 \cdot s$

<u>Theorem 3</u>

A piece wise smooth weak solution of a homogeneous polynomial nonlinear partial differential equation is a regular weak solution.

<u>Proof</u>

It follows from the relations (18), (19) and (24) in the proof of Theorem 1 in §2
▽▽▽

<u>Theorem 4</u>

Given a nonlinear partial differential operator $T(D)$ of the form in (39) and a regular weak solution $S \in \mathcal{D}'(\Omega)$ of the equation

(44) $T(D)u(x) = 0$, $x \in \Omega$,

there exist regularizations $(\mathcal{V},\mathcal{S}')$ such that for any admissible property Q , one obtains

1) $S \in A^Q(\mathcal{V},\mathcal{S}',p)$, $\forall\ p \in \bar{N}^n$

2) in the case of <u>derivative algebras,</u> S satisfies the equation (44) in the <u>usual algebraic sense</u>, with the respective multiplication and derivatives within the algebras $A^Q(\mathcal{V},\mathcal{S}',p)$, $p \in \bar{N}^n$.

<u>Proof</u>

Since S is a regular weak solution of (44), there exists $s \in \mathcal{S}_o(\Omega)$ satisfying (41), (42) and (43).

We notice that the ideal $\mathcal{I}_v$ is regular in the sense of chapter 2, §10. Indeed, the condition (43) above implies (55) in chap. 2, §10. Further, (41.1) and (42) above imply

$$v(\nu) = 0 \quad \text{on} \quad G \,, \quad \forall \ \nu \in N \,,$$

therefore

$$(45) \qquad \begin{array}{l} \forall \ w \in I_v : \\ \qquad w(\nu) = 0 \quad \text{on} \quad G \,, \quad \forall \ \nu \in N \end{array}$$

which obviously implies (56) in chap. 2, §10. Therefore, I_v is indeed a regular ideal in $W(\Omega)$.

Assume now given a regular ideal I in $W(\Omega)$, such that

(46) $\qquad I \supset I_v$

(47) $\qquad I \cap (V_0 + U + R) \subset V_0$

Then, according to Proposition 9 in chap. 2, §10, there exist vector subspaces T in S_0 compatible with I , as well as vector subspaces S_1 in S_0 satisfying

(48) $\qquad V_0 \oplus T \oplus S_1 = S_0$

(49) $\qquad U \subset S_1$

In case S is not smooth, (47) implies

(50) $\qquad s \notin V_0 + T + U$

since $V_0 + T \subset V_0 + I$, if one takes T as in the proof of Proposition 9 in chap. 2, §10. Then (48), (49) and (50) imply that S_1 can be chosen satisfying

(51) $\qquad s \in S_1$

Now, one can choose a vector subspace V in $I \cap V_0(\Omega)$ such that

(52) $\qquad v \in V(p) \,, \quad \forall \ p \in \bar{N}^n$

since $v \in I_v \subset I$. Denoting

(53) $\qquad S' = T \oplus S_1$

one obtains a regularization (V,S') according to Theorem 7 in chap. 2, §10.

Given an admissible property Q , the relation

(54) $\qquad S \in A^Q(V,S',p) \,, \quad \forall \ p \in \bar{N}^n$

results from (41) and the fact that $D'(\Omega) \subset A^Q(V,S',p) \,, \quad \forall \ p \in \bar{N}^n$.

It only remains to show that S satisfies (44) in the usual algebraic sense, with the respective multiplication and derivatives in the algebras $A^Q(V,S',p) \,, \ p \in \bar{N}^n$.

The relation (51) will give

(55) $\qquad S = s + I^Q(V(p),S') \in A^Q(V,S',p) \,, \quad \forall \ p \in \bar{N}^n$

Now, §9 and Theorems 2 and 3 in §8, chap. 1 as well as (52) above, imply in the case of derivative algebras, the relation

(56) $T(D)S = T(D)s + I^Q(V(p),S') = v + I^Q(V(p),S') = 0 \in A^Q(V,S',p) \ , \quad \forall \ p \in \bar{N}^n$

The relations (55) and (56) end the proof of Theorem 4 ∇∇∇

Remark 2

1) The condition (43) in the definition of a regular weak solution can be written under the following explicit form:

$$\forall \quad \psi \in C^\infty(\Omega) \ :$$

$$u(\psi) = v_1 + \sum_i w_i \prod_j D^{p_{ij}} T(D)s \Rightarrow \psi = 0 \quad \text{on} \quad \Omega \ ,$$

where $v_1 \in V_0(\Omega)$, $w_i \in W(\Omega)$ and $p_{ij} \in N^n$.

Taking into account Remark 3 in chap. 2, §10, the condition (43) above can be replaced by the weaker one given in (70), in the mentioned remark.

2) The algebras $A^Q(V,S',p)$ obtained in Theorem 4, are obviously regular algebras in the sense of chapter 2, §10.

Chapter 4

QUANTUM PARTICLE SCATTERING IN POTENTIALS POSITIVE POWERS OF THE DIRAC δ DISTRIBUTION

§1. INTRODUCTION

Potentials with strong local singularities have been studied in scattering theory, [3], [27], [28], [115], [116], [140]. The strongest local singularities of the potentials considered were those of measures which need not be absolutely continuous with respect to the Lebesque measure, [27]. The potentials in this chapter, given by arbitrary positive powers $(\delta)^m$, $0 < m < \infty$, of the Dirac δ distribution, present obviously stronger local singularities.

The wave function solutions obtained possess the scattering property of being given by pairs ψ_-, ψ_+ of usual C^∞ solutions of the potential free motions, each valid on the respective side of the potential and satisfying special junction relations on the support of the potentials. In the case of the potential $\alpha\delta$, i.e. $m = 1$, the only one treated in literature, [44], the junction relation obtained is identical with the known one.

§2. WAVE FUNCTIONS, JUNCTION RELATIONS

One and three dimensional motions are considered.
The one dimensional wave function ψ is given by

$$(1) \qquad \psi''(x) + (k-U(x))\psi(x) = 0, \quad x \in R^1 \quad (k \in R^1)$$

with the potential

$$(2) \qquad U(x) = \alpha(\delta(x))^m, \quad x \in R^1 \quad (\alpha \in R^1, \quad m \in (0,\infty))$$

The solution of (1), (2) is expected to be of the form

$$(3) \qquad \psi(x) = \begin{vmatrix} \psi_-(x) & \text{if } x < 0 \\ \psi_+(x) & \text{if } x > 0 \end{vmatrix}$$

where ψ_-, $\psi_+ \in C^\infty(R^1)$ are solutions of

$$\psi''(x) + k\psi(x) = 0, \quad x \in R^1,$$

satisfying certain initial conditions

$$\psi_-(x_0) = y_0, \quad \psi_-'(x_0) = y_1$$

$$\psi_+(x_1) = z_o \; , \quad \psi_+'(x_1) = z_1 \; ,$$

where $-\infty \le x_o \le 0 \le x_1 \le \infty$ and y_o , y_1 , z_o , $z_1 \in C^1$ are given and the vectors

$$\begin{pmatrix} y_o \\ y_1 \end{pmatrix} , \begin{pmatrix} z_o \\ z_1 \end{pmatrix}$$

might be in a certain relation.

As known, [44], that is the situation in the case of $m = 1$ and $x_o = x_1 = 0$, when the junction relation in $x = 0$ between ψ_- and ψ_+ is given by

$$(4) \qquad \begin{pmatrix} \psi_+(0) \\ \psi_+'(0) \end{pmatrix} = \begin{pmatrix} 1 & 0 \\ \alpha & 1 \end{pmatrix} \begin{pmatrix} \psi_-(0) \\ \psi_-'(0) \end{pmatrix}$$

In the case of an arbitrary positive power $m \in (0,\infty)$, the following three problems arise:

1) to define the power $(\delta(x))^m$, $x \in R^1$, of the Dirac δ distribution,
2) to prove that the hypothesis (3) is correct, and
3) to obtain a junction relation generalizing (4).

The first problem is solved in §5, where a special case of the Dirac algebras constructed in chapter 2 will be employed. The solution of the second problem results from Theorem 4 in §5, and is based on the smooth representation of δ constructed in §4. The third problem will be the one solved first, using a standard 'weak solution' approach presented in §3. That approach will also suggest the way the first two problems can be solved.

The junction relations in $x = 0$ between ψ_- and ψ_+ , will be:

$$(5) \qquad \begin{pmatrix} \psi_+(0) \\ \psi_+'(0) \end{pmatrix} = Z(m,\alpha) \begin{pmatrix} \psi_-(0) \\ \psi_-'(0) \end{pmatrix}$$

where

$$(5.1) \qquad Z(m,\alpha) = \begin{pmatrix} 1 & 0 \\ 0 & 1 \end{pmatrix} , \quad \text{for } m \in (0,1) , \; \alpha \in R^1 ,$$

$$(5.2) \qquad Z(1,\alpha) = \begin{pmatrix} 1 & 0 \\ \alpha & 1 \end{pmatrix} , \quad \text{for } \alpha \in R^1 \quad \text{(see [44] and (4))}$$

$$(5.3) \qquad Z(2, -(\nu\pi)^2) = \begin{pmatrix} (-1)^\nu & 0 \\ 0 & (-1)^\nu \end{pmatrix} , \quad \text{for } \nu = 0,1,2,\ldots$$

$$(5.4) \qquad Z(m,\alpha) = \begin{pmatrix} \sigma & 0 \\ K & \sigma \end{pmatrix} , \quad \text{for } m \in (2,\infty) , \; \alpha \in (-\infty,0)$$

with $\sigma = \pm 1$ and $-\infty \le K \le +\infty$ arbitrary.

The interpretation of (5) in the case of one dimentional motions (1) in potentials (2) results as follows:

1) For $m = 1$, the known, [44], motion is obtained.

2) If $m = 2$, then for the discrete levels of the potential well

(6) $$U(x) = -(\nu\pi)^2(\delta(x))^2 \ , \quad x \in R^1 \ , \quad \nu = 1,3,5,7,\ldots$$

there is motion through the potential, which causes a sign change of the wave function, namely $\psi_+(x) = -\psi_-(x)$, $x \in R^1$.

3) If $m \in (2,\infty)$, there is motion through the potential (2) in the case of a potential well only; however, the junction relation (5.4) will not give a unique connection in $x = 0$ between ψ_- and ψ_+ as the parameters σ and K involved can be arbitrary.

As known, [44], the problem of the three dimensional spherically symmetric motion with no angular momentum, and the radial wave function R given by

$$(r^2R'(r))' + r^2(k-U(r)) \cdot R(r) = 0 \ , \quad r \in (0,\infty) \quad (k \in R^1)$$

where the potential concentrated on the sphere of radius a is

$$U(r) = \alpha(\delta(r-a))^m \ , \quad r \in (0,\infty) \quad (\alpha \in R^1 \ , m \ , a \in (0,\infty)) \ ,$$

can be reduced to the solution of (1), (2). Therefore, the above interpretation for the one dimensional motion will lead to the corresponding interpretation for the three dimensional motion.

§3. WEAK SOLUTION

The solution (3), (5) of (1), (2) will be obtained in two steps.
First, a convenient nonsmooth representation of δ will give in Theorem 1 a weak solution of (1), (2).

The second step, in §4, constructs a smooth representation of δ, needed in the algebras containing $\mathcal{D}'(R^1)$. That representation gives the same weak solution, which proves to be a valid solution of (1), (2) within the mentioned algebras and therefore, independent of the representations used for δ.

The nonsmooth representation of δ, employed for the sake of simpler computation of the junction relations, is given in

(7) $$\delta(x) = \lim_{\nu\to\infty} V(\omega_\nu \ , 1/\omega_\nu \ , x) \ , \quad x \in R^1 \ , \quad \nu \in N \ ,$$

where

(8) $$\lim_{\nu\to\infty} \omega_\nu = 0 \quad \text{and} \quad \omega_\nu > 0 \quad \text{with} \quad \nu \in N \ ,$$

while

$$(9) \qquad V(\omega,K,x) = \begin{cases} K & \text{if} \quad 0 < x < \omega \\ 0 & \text{if} \quad x \leq 0 \quad \text{or} \quad x \geq \omega \end{cases}$$

where $\omega > 0$ and $K \in R^1$.

Given $m \in (0,\infty), \alpha, k \in R^1$, $x_o < 0$, $y_o, y_1 \in C^1$ and $\nu \in N$, denote by $\psi_\nu \in C^\infty(R^1 \setminus \{0,\omega_\nu\}) \cap C^1(R^1)$, the unique solution of

$$(10) \qquad \psi''(x) + (k - V(\omega_\nu, \alpha/(\omega_\nu)^m, x))\psi(x) = 0 , \quad x \in R^1 ,$$

with the initial conditions

$$(11) \qquad \psi(x_o) = y_o , \quad \psi'(x_o) = y_1$$

Denote by $M(k,x_o)$ the set of all $(m,\alpha) \in (0,\infty) \times R^1$ for which there exists $(\omega_\nu \mid \nu \in N)$ satisfying (8) and such that

$$(12) \qquad \lim_{\nu \to \infty} \begin{pmatrix} \psi_\nu(\omega_\nu) \\ \psi'_\nu(\omega_\nu) \end{pmatrix} = \begin{pmatrix} z_o \\ z_1 \end{pmatrix}$$ exists and finite, for any $y_o, y_1 \in C^1$.

Suppose given $(m,\alpha) \in M(k,x_o)$ and $(\omega_\nu \mid \nu \in N)$ satisfying (8) and (12). Then for any $y_o, y_1 \in C^1$, one can define $\psi_-, \psi_+ \in C^\infty(R^1)$ as the unique solutions of

$$(13) \qquad \psi''(x) + k\psi(x) = 0 , \quad x \in R^1 ,$$

satisfying respectively the initial conditions

$$(14) \qquad \begin{pmatrix} \psi_-(x_o) \\ \psi'_-(x_o) \end{pmatrix} = \begin{pmatrix} y_o \\ y_1 \end{pmatrix} , \quad \begin{pmatrix} \psi_+(0) \\ \psi'_+(0) \end{pmatrix} = \begin{pmatrix} z_o \\ z_1 \end{pmatrix}$$

where $z_o, z_1 \in C^1$ is obtained through (12).

Theorem 1

Suppose ψ given in (3) with ψ_-, ψ_+ from (13) and (14). Then, the sequence of functions $(\psi_\nu \mid \nu \in N)$ resulting from (10) and (11) is convergent in $D'(R^1)$ to ψ.

Proof

Obviously $\psi_\nu = \psi_-$ on $[-\infty,0]$, for every $\nu \in N$. Thus, it remains to evaluate $\psi_+ - \psi_\nu$ on $(0,\infty)$. The relation (12) and the second relation in (14) imply that

$$(15) \qquad \begin{aligned} &\forall \; a,\varepsilon > 0 \; : \; \exists \; \mu \in N \; : \; \forall \; \nu \in N , \; \nu \geq \mu \; : \\ &\quad | \psi_+ - \psi_\nu | , \; | \psi'_+ - \psi'_\nu | \leq \varepsilon \quad \text{on } [\omega_\nu , a] \end{aligned}$$

Now, from the proof of Theorem 2, below, on can obtain that

$$(16)\qquad \begin{array}{l} \exists\ K > 0\ :\ \forall\ \nu \in N\ : \\ |\ \psi_\nu\ |\ ,\ |\ \psi'_\nu\ | \leq K \quad \text{on} \quad [0,\omega_\nu] \end{array}$$

Indeed, according to (19) in the proof of Theorem 2, it follows that

$$\begin{pmatrix} \psi_\nu(x) \\ \psi'_\nu(x) \end{pmatrix} = W(k-\alpha/(\omega_\nu)^m,x) \begin{pmatrix} \psi_-(0) \\ \psi'_-(0) \end{pmatrix} , \quad \forall \quad \nu \in N\ ,\quad x \in [0,\omega_\nu]\ .$$

which implies the following two evaluations, respectively for $\alpha > 0$ and $\alpha < 0$.

Assume $\alpha > 0$, then for any $\nu \in N$ and $x \in [0,\omega_\nu]$, one obtaines

$$(17)\qquad \begin{array}{rl} |\ \psi_\nu(x) - \psi_\nu(\omega_\nu)\ | \leq & (\ |\exp(xH_\nu) - \exp L_\nu| + |\exp(-xH_\nu) - \exp(-L_\nu)|\ ) \cdot \\ & \cdot\ (\ |\psi_-(0)| + |\psi'_-(0)|/H_\nu\ )\ /\ 2 \ \leq \\ \leq & (\exp L_\nu + 1) \cdot (\ |\psi_-(0)| + |\psi'_-(0)|/H_\nu\ )\ /\ 2 \end{array}$$

the last inequality resulting from the fact that $0 \leq xH_\nu \leq L_\nu$ since $0 \leq x \leq \omega_\nu$. Now, the relations (21) and (22) in the proof of Theorem 2, together with (17) and (12) will imply (16).

Assume $\alpha < 0$, then for any $\nu \in N$ and $x \in [0,\omega_\nu]$, one obtaines

$$(18)\qquad \begin{array}{rl} |\ \psi_\nu(x) - \psi_\nu(\omega_\nu)\ | \leq & |\ \cos xH_\nu - \cos L_\nu\ | \cdot |\ \psi_-(0)\ | + \\ & +\ |\ \sin xH_\nu - \sin L_\nu\ | \cdot |\ \psi'_-(0)\ |\ /\ H_\nu \end{array}$$

Now, the relation (23) in the proof of Theorem 2, together with (18) and (12), will again imply (16). The relations (15) and (16) obviously complete the proof ▽▽▽

According to Theorem 1, the function ψ in (3) with ψ_- , ψ_+ from (13), (14) is a weak solution of (1), (2) obtained by the respresentation of δ in (7), (8), (9), provided the potential (2) is obtained from $(m,\alpha) \in M(k,x_0)$. The problem of the structure of $M(k,x_0)$ is solved now.

<u>Theorem 2</u>

The set $M(k,x_0)$ does not depend on $k \in R^1$ and $x_0 < 0$, and

$$M = ((0,1] \times R^1) \cup (\{2\} \times \{-\pi^2,-4\pi^2,-9\pi^2,\ldots\}) \cup ((2,\infty) \times (-\infty,0)) \cup \cup ((0,\infty) \times \{0\})$$

<u>Proof</u>

If $u \in C^\infty(R^1)$ is the unique solution of

$$u''(x) + h\,u(x) = 0\ , \quad x \in R^1\ , \quad (h \in R^1)$$

with the initial conditions

$$u(a) = b\ , \quad u'(a) = c\ ,$$

then

$$(19)\qquad \begin{pmatrix} u(x) \\ u'(x) \end{pmatrix} = W(h,x)\ W(h,-a) \begin{pmatrix} b \\ c \end{pmatrix} , \quad x \in R^1\ ,$$

where

$$W(h,x) = \exp(xA_h) \quad , \quad A_h = \begin{pmatrix} 0 & 1 \\ -h & 0 \end{pmatrix}$$

Assume $(m,\alpha) \in (0,\infty) \times R^1$. Applying (19) to the functions ψ_ν , one obtains

$$\begin{pmatrix} \psi_\nu(\omega_\nu) \\ \psi_\nu'(\omega_\nu) \end{pmatrix} = W\,(k-\alpha/(\omega_\nu)^m,\omega_\nu)\; W\,(k,-x_o) \begin{pmatrix} y_o \\ y_1 \end{pmatrix} \quad , \quad \forall \quad \nu \in N$$

Therefore, $(m,\alpha) \in M(k,x_o)$ only if

(20) $\qquad \lim\limits_{\nu\to\infty} W(k-\alpha/(\omega_\nu)^m,\omega_\nu) = Z(m,\alpha)$ exists and finite.

It thus remains to make the condition (20) explicit in terms of m and α .

First, suppose $\alpha > 0$. Since $\omega_\nu \to 0$, one can assume $k - \alpha / (\omega_\nu)^m < 0$, therefore

$$W(k-\alpha/(\omega_\nu)^m,\omega_\nu) = \frac{1}{2} \begin{pmatrix} \exp L_\nu + \exp(-L_\nu) & \frac{1}{H_\nu}(\exp L_\nu - \exp(-L_\nu)) \\ H_\nu\,(\exp L_\nu - \exp(-L_\nu)) & \exp L_\nu + \exp(-L_\nu) \end{pmatrix}$$

with

$$H_\nu = (-k+\alpha/(\omega_\nu)^m)^{1/2} \quad , \quad L_\nu = \omega_\nu H_\nu$$

Obviously

(21) $\qquad \lim\limits_{\nu\to\infty} H_\nu = +\infty$

(22) $\qquad \lim\limits_{\nu\to\infty} L_\nu + \left| \begin{array}{lll} 0 & \text{if} & m \in (0,2) \\ \alpha^{1/2} & \text{if} & m = 2 \\ +\infty & \text{if} & m \in (2,\infty) \end{array} \right.$

Therefore, $M(k,x_o) \cap ([2,\infty) \times (0,\infty)) = \emptyset$.

Assume now $m \in (0,2)$, then the three terms in $W(k-\alpha/(\omega_\nu)^m,\omega_\nu)$, except $\frac{H_\nu}{2}(\exp L_\nu - \exp(-L_\nu))$, have got a finite limit when $\nu \to \infty$. Concerning the latter term, one obtains

$$\lim_{\nu\to\infty} \frac{H_\nu}{2}(\exp L_\nu - \exp(-L_\nu)) = \left| \begin{array}{lll} 0 & \text{if} & m \in (0,1) \\ \alpha & \text{if} & m = 1 \\ +\infty & \text{if} & m \in (1,2) \end{array} \right.$$

Thus, one can conclude that $M(k,x_o) \cap ((1,2) \times (0,\infty)) = \emptyset$ and $(0,1] \times (0,\infty) \subset M(k,x_o)$

Suppose now $\alpha < 0$. Since $\omega_\nu \to 0$, one can assume $k - \alpha / (\omega_\nu)^m > 0$, therefore

$$W(k-\alpha/(\omega_\nu)^m,\omega_\nu) = \begin{pmatrix} \cos L_\nu & \frac{1}{H_\nu} \sin L_\nu \\ -H_\nu \sin L_\nu & \cos L_\nu \end{pmatrix}$$

with

$$H_\nu = (k-\alpha/(\omega_\nu)^m)^{1/2} \quad , \quad L_\nu = \omega_\nu H_\nu$$

Obviously

$$\lim_{\nu\to\infty} H_\nu = +\infty \tag{23}$$

$$\lim_{\nu\to\infty} L_\nu = \begin{cases} 0 & \text{if } m \in (0,2) \\ (-\alpha)^{1/2} & \text{if } m = 2 \\ +\infty & \text{if } m \in (2,\infty) \end{cases}$$

Assume now $m \in (0,2)$, then the three terms in $W(k-\alpha/(\omega_\nu)^m,\omega_\nu)$, except $-H_\nu \sin L_\nu$, have got a finite limit when $\nu \to \infty$. Concerning the latter term, one obtains

$$\lim_{\nu\to\infty} (-H_\nu \sin L_\nu) = \begin{cases} 0 & \text{if } m \in (0,1) \\ \alpha & \text{if } m = 1 \\ -\infty & \text{if } m \in (1,2) \end{cases}$$

Therefore $M(k,x_0) \cap ((1,2) \times (-\infty,0)) = \emptyset$ and $(0,1] \times (-\infty,0) \subset M(k,x_0)$.

Now, assume $m = 2$, then again the three terms in $W(k-\alpha/(\omega_\nu)^m,\omega_\nu)$, except $-H_\nu \sin L_\nu$ have got a finite limit when $\nu \to \infty$, while the latter term tends to a limit according to

$$\lim_{\nu\to\infty} (-H_\nu \sin L_\nu) = \begin{cases} 0 & \text{if } \alpha = -(\mu\pi)^2 \text{ with } \mu = 1,2,\ldots \\ +\infty & \text{otherwise} \end{cases}$$

Therefore $M(k,x_0) \cap (\{2\} \times (-\infty,0)) = \{2\} \times \{-(\mu\pi)^2 \mid \mu = 1,2,\ldots\}$.

Finally, assume $m \in (2,\infty)$, then $\lim_{\nu\to\infty} H_\nu = \lim_{\nu\to\infty} L_\nu = +\infty$ thus a necessary condition for (20) is

$$\lim_{\nu\to\infty} \sin L_\nu = 0 \tag{24}$$

The condition (24) will indicate the way the sequence $(\omega_\nu \mid \nu \in N)$ satisfying (8) has to be chosen in order to secure (20). Indeed, given $k \in R^1$, there exist $A,B \in (0,\infty)$ such that $k - \alpha/\omega^m > 0$, $\forall\ \omega \in (0,A)$ and the function $\theta : (0,A) \to (B,\infty)$ defined by $\theta(\omega) = \omega(k - \alpha/\omega^m)^{1/2}$ has the properties

θ is strictly decreasing on $(0,A)$,

$\lim_{\omega\to 0} \theta(\omega) = \infty$, $\lim_{\omega\to A} \theta(\omega) = B$.

Therefore, the inverse function $\theta^{-1} : (B,\infty) \to (0,A)$ exists, is strictly decreasing on (B,∞) and

$$\lim_{\gamma\to\infty} \theta^{-1}(\gamma) = 0 \tag{25}$$

Moreover, $\theta^{-1} \in C^1(B,\infty)$ and

$$\lim_{\gamma\to\infty} D\theta^{-1}(\gamma) = 0 \tag{26}$$

Assume now that $(n_\nu \mid \nu \in N)$ is a sequence of positive integers and $(e_\nu \mid \nu \in N)$ is a sequence of nonzero real numbers, such that

$$\lim_{\nu\to\infty} n_\nu = \infty, \quad \lim_{\nu\to\infty} e_\nu = 0 \quad \text{and} \quad n_\nu\pi + e_\nu > B, \quad \forall\ \nu \in N. \tag{27}$$

Define

(28) $\omega_\nu = \theta^{-1}(n_\nu \pi + e_\nu) \quad , \quad \forall \ \nu \in N$

Then $(\omega_\nu \mid \nu \in N)$ satisfies (8), according to (27) and (25). Further, one obtains

(29) $\cos L_\nu = (-1)^{n_\nu} \cos e_\nu \ , \ -H_\nu \sin L_\nu = (-1)^{n_\nu+1} H_\nu \sin e_\nu \ , \quad \forall \ \nu \in N$

Now, (29) and (27) imply that

(30) $\lim_{\nu\to\infty} \cos L_\nu$ exists

provided that

(31) n_ν , with $\nu \in N$, have constant parity.

Therefore, (20) will hold only if

(32) $\lim_{\nu\to\infty} (-H_\nu \sin L_\nu)$ exists and finite

But, due to (29), (27) and (31), the property in (32) is equivalent with

(33) $\lim_{\nu\to\infty} e_\nu H_\nu$ exists and finite

It is simpler to compute the square of the limit in (33) which due to (28), (27) and (26) becomes

$$\lim_{\nu\to\infty} (e_\nu H_\nu)^2 = \lim_{\nu\to\infty} (e_\nu)^2 (k - \alpha/(\theta^{-1}(n_\nu \pi + e_\nu))^m) =$$
$$= -\alpha \lim_{\nu\to\infty} (e_\nu)^2 \ / \ (\theta^{-1}(n_\nu \pi + e_\nu))^m =$$
$$= -\alpha \lim_{\nu\to\infty} (\ |e_\nu|^{2/m} \ \theta^{-1}(n_\nu \pi) + |e_\nu|^{1-2/m} \ D\theta^{-1}(n_\nu \pi + \xi_\nu e_\nu))^{-m} =$$
$$= -\alpha \lim_{\nu\to\infty} (e_\nu)^2 \ / \ (\theta^{-1}(n_\nu \pi))^m$$

since $\xi_\nu \in (0,1)$, $\forall \ \nu \in N$.

But, due to (25) and (27), the last limit can assume any value in $[0,\infty]$, depending on a proper choice of n_ν and e_ν . Therefore, (30) and the second relation in (29) will imply that for any $\sigma \in \{-1,1\}$ and $K \in [-\infty , +\infty]$, there exists $(\omega_\nu \mid \nu \in N)$ satis fying (8) and such that

$$\lim_{\nu\to\infty} W(k - \alpha/(\omega_\nu)^m, \omega_\nu) = \begin{pmatrix} \sigma & 0 \\ K & \sigma \end{pmatrix}$$

Now, obviously $(2,\infty) \times (-\infty,0) \subset \mathcal{M}(k,x_0)$ and the proof is completed ∇∇∇

Remark 1

The relations (5.1) - (5.4) result easily from the proof of Theorem 2.

§4. SMOOTH REPRESENTATIONS FOR δ

In order to prove that the weak solutions (3), (5) of (1), (2) obtained in §2, are valid within the algebras containing the distributions and therefore, independent of the representations (7), (8), (9) used for δ, we first need to show that the same weak solutions can be obtained from certain smooth representations of δ. These representations will be obtained by appropriately 'rounding off the corners' in (7), (8) and (9). The 'rounding off' is accomplished with the help of any pair of functions $\beta, \gamma \in C^{\infty}_{+}(R^1)$ (see chap. 1, §8) satisfying:

(34)
$$\begin{aligned} &*) \quad \beta = 0 \quad \text{on} \quad (-\infty,-1] \\ &**) \quad 0 \le \beta \le M \quad \text{on} \quad (-1,1) \\ &***) \quad \beta = 1 \quad \text{on} \quad [1,\infty) \\ &****) \quad D^p\beta(0) \neq 0 \ , \quad \forall \ p \in N \end{aligned}$$

and

(35)
$$\begin{aligned} &*) \quad \gamma = 1 \quad \text{on} \quad (-\infty,-1] \\ &**) \quad 0 \le \gamma \le 1 \quad \text{on} \quad (-1,1) \\ &***) \quad \gamma = 0 \quad \text{on} \quad [1,\infty) \end{aligned}$$

The existence of the functions β and γ results from Lemma 1, at the end of this section.

Given now a sequence $(\omega_\nu \mid \nu \in N)$ satisfying (8) and two other sequences $(\omega'_\nu \mid \nu \in N)$, $(\omega''_\nu \mid \nu \in N)$ such that

(36)
$$\omega'_\nu \ , \ \omega''_\nu > 0 \ , \quad \forall \ \nu \in N \ , \quad \omega'_0 \ , \ \omega'_1 \ , \ \omega'_2 \ , \ \ldots \text{ are pair wise different and}$$
$$\lim_{\nu\to\infty} (\omega'_\nu + \omega''_\nu) \ / \ \omega_\nu = 0 \ ,$$

define $s_\delta \in W$ by

(37)
$$s_\delta(\nu)(x) = \beta(x/\omega'_\nu)\ \gamma\ ((x-\omega_\nu)/\omega''_\nu) \ / \ \omega_\nu \ , \qquad \forall \ \nu \in N \ , \quad x \in R^1 \ ,$$

then

$$\operatorname{supp} s_\delta(\nu) \subset [-\omega'_\nu \ , \ \omega_\nu + \omega''_\nu] \quad \text{and} \quad \left|1 - \int_{R^1} s_\delta(\nu)(x)dx\right| \le 2((M+1)\omega'_\nu + \omega''_\nu)/\omega_\nu \ ,$$
$$\forall \ \nu \in N \ ,$$

therefore, due to (36), one obtains

(38)
$$s_\delta \in S_0 \cap W_+ \quad \text{and} \quad \langle s_\delta \ , \ \cdot \rangle = \delta$$

with the relation $s_\delta \in W_+$ (see chap. 1, §8) implied by (37) and the fact that $\beta, \gamma \in C^{\infty}_{+}(R^1)$.

Now, the smooth representation of δ obtained in (37) will be the one replacing (7), (8) and (9). It remains to prove that (37) generates again the weak solution (3), (5) when used in solving (1), (2). Given $(m,\alpha) \in M$, $k \in R^1$, $x_0 < 0$, $(\omega_\nu \mid \nu \in N)$ satisfying (8) and (12), $(\omega'_\nu \mid \nu \in N)$ and $(\omega''_\nu \mid \nu \in N)$ satisfying (36), y_1 , $y_2 \in C^1$ and $\nu \in N$, denote by $\chi_\nu \in C^{\infty}(R^1)$ the unique solution of

(39) $$\chi''(x) + (k-\alpha(s_\delta(\nu)(x))^m)\ \chi\ (x) = 0\ , \quad x \in R^1\ ,$$

with the initial conditions

(40) $$\chi(x_0) = y_0\ ,\ \chi'(x_0) = y_1$$

Theorem 3

It is possible to choose $(\omega'_\nu \mid \nu \in N)$ and $(\omega''_\nu \mid \nu \in N)$ satisfying (36), and such that the sequence of functions $(\chi_\nu \mid \nu \in N)$ resulting from (39) and (40) is convergent in $\mathcal{D}'(R^1)$ to ψ given in (3), where ψ_- and ψ_+ are from (13) and (14).

Proof

A Gronwall inequality argument will be used. First, the equations (10), (11) are written under the form

$$F'_\nu(x) = A_\nu(x)\ F_\nu(x)\ , \quad x \in R^1\ , \quad F_\nu(x_0) = \begin{pmatrix} y_0 \\ y_1 \end{pmatrix}$$

where

$$F_\nu(x) = \begin{pmatrix} \psi_\nu(x) \\ \psi'_\nu(x) \end{pmatrix}\ , \quad A_\nu(x) = \begin{pmatrix} 0 & 1 \\ -k+V(\omega_\nu\ ,\ \alpha/(\omega_\nu)^m, x) & 0 \end{pmatrix}$$

Similarly, (39), (40) can be written as

$$G'_\nu(x) = B_\nu(x)\ G_\nu(x)\ , \quad x \in R^1\ , \quad G_\nu(x_0) = \begin{pmatrix} y_0 \\ y_1 \end{pmatrix}$$

with

$$G_\nu(x) = \begin{pmatrix} \chi_\nu(x) \\ \chi'_\nu(x) \end{pmatrix}\ , \quad B_\nu(x) = \begin{pmatrix} 0 & 1 \\ -k+\alpha(s_\delta(\nu)(x))^m & 0 \end{pmatrix}$$

Denote $H_\nu = F_\nu - G_\nu$, then

$$H'_\nu(x) = B_\nu(x)\ H_\nu(x) + (A_\nu(x)-B_\nu(x))\ F_\nu(x)\ , \quad x \in R^1\ , \quad H_\nu(x_0) = \begin{pmatrix} 0 \\ 0 \end{pmatrix}$$

therefore

$$H_\nu(x) = \int_{x_0}^{x} (A_\nu(\xi) - B_\nu(\xi))\ F_\nu(\xi)d\xi + \int_{x_0}^{x} B_\nu(\xi)\ H_\nu(\xi)d\xi\ , \quad x \in R^1$$

Applying the $||\ \ ||_\infty$ vector, respectively matrix norms, denoted for simplicity by $||\ \ ||$, one obtains

$$||H_\nu(x)|| \le \int_{x_0}^{x} ||A_\nu(\xi)-B_\nu(\xi)|| \cdot ||F_\nu(\xi)||\ d\xi + \int_{x_0}^{x} ||B_\nu(\xi)||\cdot||H_\nu(\xi)||\ d\xi\ , \quad x \in R^1\ .$$

Now, the Gronwall inequality implies

$$||H_\nu(x)|| \le \int_{x_0}^{x} ||A_\nu(\xi)-B_\nu(\xi)|| \cdot ||F_\nu(\xi)|| \cdot (\exp \int_{\xi}^{x} ||B_\nu(\eta)||\, d\eta)\, d\xi \,, \quad x \in R^1$$

But

$$|\ \psi_\nu(x) - \chi_\nu(x)\ | \le ||H_\nu(x)|| \,, \quad x \in R^1$$

and

$$||A_\nu(\xi)-B_\nu(\xi)|| = 0 \,, \quad \xi \in (-\infty,-\omega_\nu'] \cup [\omega_\nu' \,, \omega_\nu-\omega_\nu''] \cup [\omega_\nu+\omega_\nu'' \,, \infty)$$

while

$$||A_\nu(\xi)-B_\nu(\xi)|| \le |\ \alpha\ | \,/\, (\omega_\nu)^m \,, \quad \xi \in [-\omega_\nu' \,, \omega_\nu'] \cup [\omega_\nu-\omega_\nu'' \,, \omega_\nu+\omega_\nu'']$$

Further

$$||B_\nu(\eta)|| \le \max \{ 1 \,, |\ k\ | + |\ \alpha\ | \,/\, (\omega_\nu)^m \} \,, \quad \eta \in R^1 \,.$$

Therefore, one obtains

$$(41) \qquad |\psi_\nu(x)-\chi_\nu(x)| \le 2(\omega_\nu'+\omega_\nu'') \cdot |\alpha| \cdot K_\nu \cdot \exp(2(\omega_\nu'+\omega_\nu'')(1+|k|+|\alpha|/(\omega_\nu)^m))/(\omega_\nu)^m \,, \quad x \in R^1$$

where

$$K_\nu = \max \{ ||F_\nu(\xi)|| \mid \xi \in [-\omega_\nu' \,, \omega_\nu'] \cup [\omega_\nu-\omega_\nu'' \,, \omega_\nu+\omega_\nu''] \}$$

For given $\nu \in N$, F_ν depends only on ω_ν and not on ω_ν' or ω_ν'' . Therefore, K_ν is decreasing in ω_ν' and ω_ν'' . That fact, together with (41) imply that for given $\nu \in N$ and $\omega_\nu > 0$, the function ψ_ν can be arbitrarily and in a uniform way on R^1 approximated by the function χ_ν , provided ω_ν' and ω_ν'' are chosen small enough. Taking into account Theorem 1 in § 3, the proof is completed ∇∇∇

<u>Lemma 1</u>

There exist functions $\beta, \gamma \in C_+^\infty(R^1)$ satisfying (34) and (35) respectively.

<u>Proof</u>

Define $\eta \in C_+^\infty(R^1)$ by

$$\eta(x) = \begin{cases} 0 & \text{if } x \le 0 \\ \exp(-1/x) & \text{if } x > 0 \end{cases}$$

Assume $0 < a \,, b < 1$ and define $\beta_1 \,, \beta_2 \in C^\infty(R^1)$ by $\beta_1(x) = \eta(x+1) \,/\, (\eta(x+1) + \eta(-x-a))$ and $\beta_2(x) = \eta(1-x) \,/\, (\eta(1-x) + \eta(x-b))$, for $x \in R^1$. Defining $\beta \in C^\infty(R^1)$ by (see Fig. 3) $\beta(x) = (\beta_1(x) \exp x-1)\, \beta_2(x) + 1$, for $x \in R^1$, β will satisfy (34) with $M = e$. Defining $\gamma \in C^\infty(R^1)$ by $\gamma(x) = \eta(1-x) \,/\, (\eta(1-x) + \eta(x+1))$, for $x \in R^1$, γ will satisfy (35) ∇∇∇

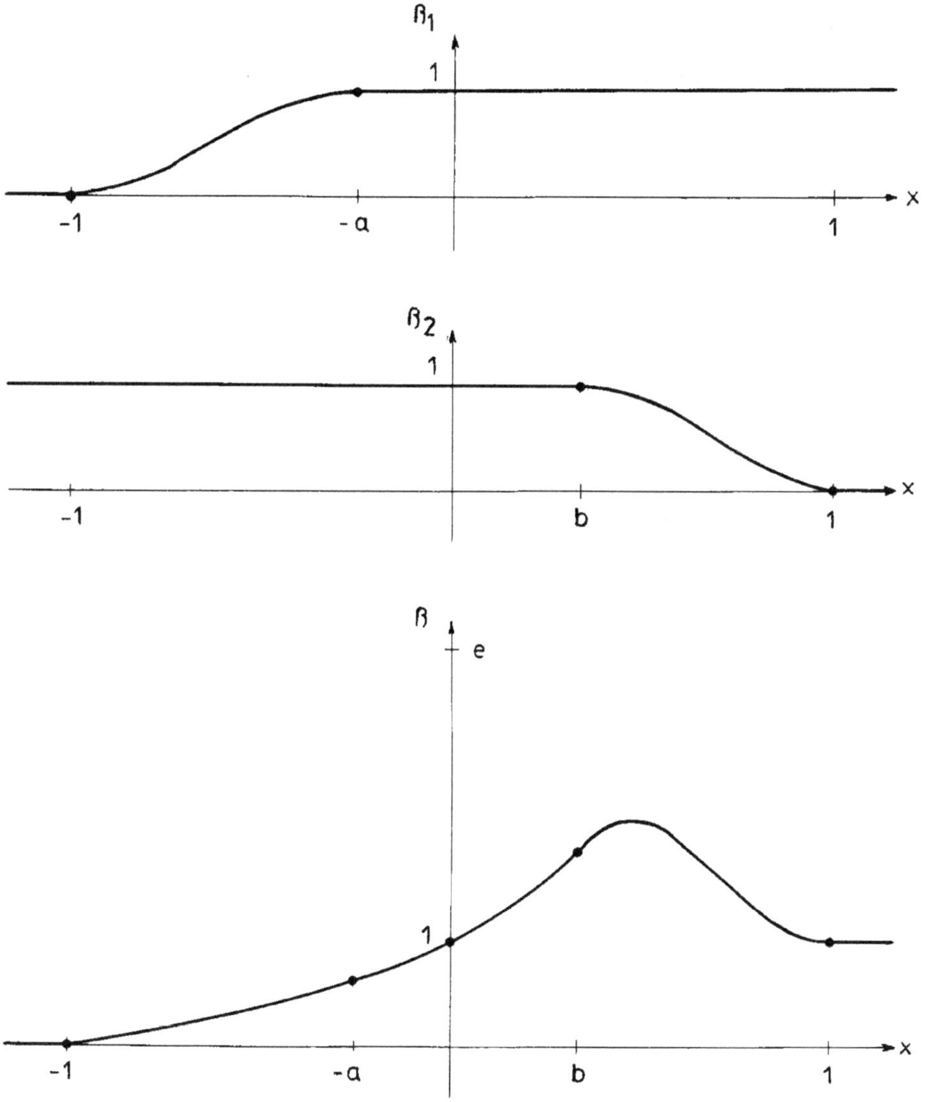

Fig. 3

§5. WAVE FUNCTION SOLUTIONS IN THE ALGEBRAS CONTAINING THE DISTRIBUTIONS

It is shown in this section that, given any $(m,\alpha) \in M$, the weak solution ψ of (1), (2) obtained in §§3, 4 is a solution of (1), (2) in a usual algebraic sense, considered in certain algebras containing $D'(R^1)$, with the multiplication, derivatives and positive powers defined in the algebras. Therefore, the wave function solution ψ obtained is independent of the particular representations used for the Dirac δ distribution.

Theorem 4

Suppose given (1), (2) with $(m,\alpha) \in M$ and let ψ be the weak solution of (1), (2) constructed in §§3, 4.

Suppose ψ is not smooth, that is, ψ or ψ' is not continuous in $0 \in R^1$. Then, there exist regularizations (V,S') (see chap. 1, §7) such that for any admissible property Q, one obtains

1) $\psi \in A^Q(V,S',p)$, $\forall\ p \in \bar{N}$

2) in the case of derivative and positive power algebras (see chap. 1, §7) ψ satisfies (1) in the usual algebraic sense in each of the algebras $A^Q(V,S',p)$, $p \in \bar{N}$, with the respective multiplication, power and derivatives.

Moreover, there exist $s \in S_0$ not depending on Q or p, such that

3) $\psi = \langle s,\cdot\rangle = s + I^Q(V(p),S') \in A^Q(V,S',p)$, $\forall\ p \in \bar{N}$.

Proof

Since $(m,\alpha) \in M$, there exists $(\omega_\nu \mid \nu \in N)$ satisfying (8) and (12). Assume given $x_0 < 0$ and $y_0, y_1 \in C^1$, then according to Theorem 3 in §4, it is possible to choose $(\omega'_\nu \mid \nu \in N)$ and $(\omega''_\nu \mid \nu \in N)$ satisfying (36) and so that the sequence of smooth functions $(\chi_\nu \mid \nu \in N)$ resulting from (39) and (40) will converge in $D'(R^1)$ to ψ given through (3), (13), (14). Therefore, defining $s \in W$ by $s(\nu) = \chi_\nu$, $\forall\ \nu \in N$, one obtains

(42) $$s \in S_0 , \quad \langle s,\cdot\rangle = \psi$$

and, due to (39), (40), the relation

(43) $$D^2 s + (k-\alpha(s_\delta)^m)s = u(0) \in 0$$

and

(44) $$\begin{pmatrix} s(\nu)(x_0) \\ Ds(\nu)(x_0) \end{pmatrix} = \begin{pmatrix} y_0 \\ y_1 \end{pmatrix} , \quad \forall\ \nu \in N .$$

The idea of the proof is to show that (43) is valid in the algebras $A^Q(V,S',p)$, with suitably chosen regularizations (V,S') . In this respect it suffices that the regularization (V,S') satisfies the condition:

(45) $\quad s, s_\delta \in V(p) \oplus S' \ , \quad \forall \ p \in \bar{N}$ (see (22) in chap. 1, §7)

Indeed, $s \in V(p) \oplus S'$, $\forall \ p \in \bar{N}$ and (42) imply (see (25) in chap. 1, §8) that

(46) $\quad \psi = s + I^Q(V(p),S') \in A^Q(V,S',p) \ , \quad \forall \ p \in \bar{N}$

In the same time, $s_\delta \in V(p) \oplus S'$, $\forall \ p \in \bar{N}$ and (38) result in

(47) $\quad \delta = s_\delta + I^Q(V(p),S') \in A^Q(V,S',p) \ , \quad \forall \ p \in \bar{N}$.

Now, (46), (47), (43), (38) and Theorems 3, 4 in chap. 1, §8, will obviously imply 2). Further, 1) and 3) will result from (46).

Therefore, it only remains to obtain regularizations (V,S') which fulfil (45).

We shall use the method given in Theorem 1, chap. 2, §3.

Take $J \subset W$ such that for each $w \in J$, the relation holds

(48) $\quad$ supp $w(\nu)$ either is void for $\nu \in N$ big enough or shrinks to $\{0\} \subset R^1$, when $\nu \to \infty$

(49) $\quad w(\nu)(0) = 0$, for $\nu \in N$ big enough

and denote by I_1 the ideal in W generated by J .

Denote by T_1 the vector subspace in S_0 generated by

$$\{s_\delta \ , \ Ds_\delta \ , \ D^2 s_\delta \ , \ \dots \}$$

We prove that I_1 and T_1 are compatible. Obviously $V_0 \cap T_1 = O$. Further, $I_1 \cap S_0 \subset V_0 \oplus T_1$. Indeed, assume $t \in I_1 \cap S_0$, then (48) implies supp $\langle t, \cdot \rangle \subset \{0\}$, thus $t \in V_0 \oplus T_1$, taking into account (38). Finally, we prove that $I_1 \cap T_1 = O$. Assume indeed $t \in I_1 \cap T_1$ then

(50) $$t = \sum_{0 \le i \le p} \lambda_i D^i s_\delta$$

with $p \in N$, $\lambda_i \in C^1$. Now, according to (49) and (37), the above relation (50) implies

$$0 = t(\nu)(0) = \sum_{0 \le i \le p} \lambda_i D^i \beta(0) / \omega_\nu (\omega'_\nu)^i \ , \quad \text{for } \nu \in N \text{ big enough}$$

which due to (36) and ****) in (34) results in

$$\lambda_0 = \dots = \lambda_p = 0 \ .$$

Now, (50) will give $t \in 0$. Recalling the conditions (4) and (5) in chap. 2, §3, it follows that I_1 and T_1 are compatible. Obviously, $s \notin V_0 \oplus T_1$ and $(V_0 \oplus T_1) \cap U = 0$. Moreover, $s \notin V_0 \oplus T_1 \oplus U$, since ψ is not smooth and $\langle s, \cdot \rangle = \psi$ according to (42).

Assume $I \supset I_1$, with I ideal in W and $T \supset T_1$, with T vector subspace in S_0 , such that I and T are compatible and $s \notin V_0 \oplus T \oplus U$. It follows that there exist vector subspaces S_1 in S_0 , such that $V_0 \oplus T \oplus S_1 = S_0$ and $s \in S_1$, $U \subset S_1$. Assume finally V a vector subspace in $I \cap V_0$. Now, Theorem 1 in chap. 2, §3 will imply that (V,S') with $S' = T \oplus S_1$ is a regularization. Obviously, (V,S') satisfies (45), since $s \in S_1$ and $s_\delta \in T_1 \subset T$ ∇∇∇

Remark 2

1) The regularizations (V,S') whose existence is obtained in Theorem 4 result in a rather simple, constructive way. Obviously, the algebras $A^Q(V,S',p)$ are Dirac algebras.

2) The smooth representation of δ given by $s_\delta \in S_0$ in (37) has the property

(51) $$D^p s_\delta(\nu)(0) \neq 0 \ , \quad \forall \ \nu \in N \ , \ p \in N$$

which was essentially needed in the proof of Theorem 4. That property implies in particular that no symmetric representation of δ can be used.

In chapters 5 and 6, a generalization of the relation (51) to the n-dimensional case will be used for defining important classes of Dirac algebras.

C h a p t e r   5

PRODUCTS WITH DIRAC DISTRIBUTIONS

§1. INTRODUCTION

A class of relations containing products with Dirac distributions encountered in the theory of distributions is given in

(1) $$(x-x_0)^r \cdot D^q\delta_{x_0} = 0 , \quad \forall \; x_0 \in R^n , \quad q,r \in N^n , \quad r \nleq q$$

where δ_{x_0} is the Dirac δ distribution concentrated in x_0 .

The importance of the relations (1) is due to the fact that they give an upper bound of the order of singularities the Dirac distributions and their derivatives exhibit.

It is worthwhile mentioning the role played by relations of type (1) in the way derivative operators may or may not be defined on the algebras containing the distributions (see chap. 1, §8, as well as Remark D in §7).

As a first result, Theorem 1, §4, will establish relations of type (1), within a wide class of Dirac algebras constructed in the present chapter. In §4, several other types of relations involving products with Dirac distributions will be proved valid within the mentioned algebras.

A second result is obtained in Theorem 6, §5, where known formulas in Quantum Mechanics, involving irregular products with Dirac and Heisenberg distributions are proved to be valid within the Dirac algebras constructed in the present chapter.

These algebras are obtained according to the general procedure given in Theorem 1, chap. 2, §3, which presumes the existence of compatible pairs I,T where I is an ideal in W and T is a vector subspace in S_0 . The construction of such compatible pairs is given in §§2 and 3. Here the main problem is the construction in §3 of suitable vector subspaces T , whose existence is based on a rather sophisticated algebraic argument involving generalized Vandermonde determinants. The present form of the respective conjecture in Theorem 8, §7, as well as its proof was offered by R.C. King.

A third result is presented in §8. For a subclass of the Dirac algebras constructed in §§2-4, a stronger version of the relations with products given in Theorem 1 in §4, involving this time infinite sums of Dirac distribution derivatives, is obtained.

§2. THE DIRAC IDEAL $\mathcal{I}^{\delta}$

Denote by $\mathcal{I}^{\delta}$ the set of all sequences of smooth functions $w \in \mathcal{W}$ which for any $x_0 \in R^n$ satisfy the condition

(2) $\qquad w(\nu)(x_0) = 0 , \quad$ for $\nu \in N$ big enough,

and for a certain neighbourhood V of x_0, the condition

(3) $\qquad V \cap \operatorname{supp} w(\nu)$ either is void for $\nu \in N$ big enough or shrinks to $\{x_0\}$ when $\nu \to \infty$.

Proposition 1

$\mathcal{I}^{\delta}$ is a Dirac ideal (see chap. 2, §6)

Proof

A direct check of the conditions (13) and (26) in chap. 2, will end the proof. However, it will be useful to give a second proof, showing that $\mathcal{I}^{\delta}$ is acutally an ideal $\mathcal{I}_{G,o}$ where $G = F_{\Gamma_\delta}$ and Γ_δ is a certain singularity generator on R^n (see chap. 2, §§2,4). Then, Proposition 6 in chap. 2, §6 will complete the proof.

Now, the singularity generator Γ_δ is chosen as the set of mappings $\gamma_{x_0} : R^n \to R^1$, with $x_0 \in R^n$, defined by

$$\gamma_{x_0}(x) = (||x-x_0||)^2 , \quad x \in R^n ,$$

where $||\ \ ||$ is the Euclidean norm. Then, obviously $F_{\gamma_{x_0}} = \{x_0\}$, for $x_0 \in R^n$.

Now, the relation $\mathcal{I}^{\delta} = \mathcal{I}_{G,o}$ will follow easily, ending the second proof of Proposition 1 $\nabla\nabla\nabla$

§3. COMPATIBLE DIRAC CLASSES $\mathcal{T}_{\Sigma}$

First, several auxiliary notions.

For $m \in N$ denote

$$P(n,m) = \{ p = (p_1 , \dots , p_n) \in N^n \mid |p| = p_1 + \dots + p_n \leq m \}$$

and by $\ell(n,m)$ the number of elements in $P(n,m)$.

One can see that there exists a linear order $\dashv$ on N^n such that

$$N^n = \{ p(1) , p(2) , \dots \} ,$$

$$p(1) \dashv p(2) \dashv \dots ,$$

and $\qquad P(n,m) = \{ p(1) , \dots , p(\ell(n,m)) \} , \quad \forall\ m \in N .$

It is easy to notice that $\ell(n,m)$ can be obtained from the recursive relations

$$\ell(1,m) = m + 1 , \quad \forall \quad m \in N ,$$

and

$$\ell(n+1,m) = \sum_{0 \le k \le m} \ell(n,k) , \quad \forall \quad m \in N .$$

To any sequence of smooth functions $w \in W$, the following Wronskian type infinite matrix of smooth functions will be associated with the help of the above linear order $\dashv$ on N^n :

$$W(w)(x) = \begin{pmatrix} D^{p(1)}w(0)(x) \ldots\ldots & D^{p(\mu)}w(0)(x) \ldots\ldots \\ \vdots & \\ D^{p(1)}w(\nu)(x) \ldots\ldots & D^{p(\mu)}w(\nu)(x) \ldots\ldots \\ \vdots & \vdots \end{pmatrix} , \quad x \in R^n$$

Denote by M the set of all infinite vectors of complex numbers $\Lambda = (\lambda_\mu \mid \mu \in N)$ with a finite number of nonzero components λ_μ .

An infinite matrix of complex numbers $A = (a_{\nu\mu} \mid \nu,\mu \in N)$ is called <u>column wise nonsingular</u>, only if

$$\forall \quad \Lambda \in M \quad : \quad A\Lambda \in M \Rightarrow \Lambda = 0$$

And now, the definition of an important class of weakly convergent sequences of smooth functions representing Dirac δ distributions. Given $x \in R^n$, denote by Z_x the set of all weakly convergent sequences of smooth functions $s \in S_0$, satisfying the conditions

(4) $\qquad < s , \cdot > = \delta_x$

(5) $\qquad$ supp $s(\nu)$ shrinks to $\{x\}$, when $\nu \to \infty$,

(6) $\qquad W(s)(x)$ is column wise nonsingular.

The existence of sequences $s \in Z_x$ will be proved in §7.

The condition (6) can be called, [128], <u>strong local presence</u> of the sequence s in x_0 , due to its meaning in the following particular case. Suppose, we are in the one dimensional case $n = 1$ and $\psi \in D(R^1)$ such that $\int_{R^1} \psi(x)dx = 1$. Define $s_\psi \in W$ by

(7) $\qquad s_\psi(\nu)(x) = (\nu+1)\psi(x/(\nu+1)) , \quad \forall \quad \nu \in N , \quad x \in R^1 .$

Then, s_ψ obviously satisfies (4) and (5), with $x_0 = 0 \in R^1$. Now, one can see that s_ψ will satisfy (6), only if

(8) $D^p\psi(0) \neq 0 \;, \quad \forall \; p \in N \;,$

in particular, ψ must be nonsymmetric (see (37) and ****) in (34) in chap. 4)

Denote

$$Z_\delta = \prod_{x \in R^n} Z_x$$

and for $\Sigma = (s_x \mid x \in R^n) \in Z_\delta$ denote by T_Σ the vector subspace in S_o generated by $\{D^p s_x \mid x \in R^n \;, \; p \in N^n\}$.

Proposition 2

For each $\Sigma \in Z_\delta$, T_Σ is a Dirac class which is compatible with the Dirac ideal I^δ (see chap. 2, §§3,6).

Proof

First, we prove that T_Σ is a Dirac class, that is, it satisfies (17.1), (17.2) and (24) in chap. 2. Indeed, the conditions (17.1) and (24) result easily. Assume now $t \in T_\Sigma$, $t \notin 0$, then

(9) $$t = \sum_{x \in X} \sum_{\substack{q \in N^n \\ q \leq p_x}} \lambda_{xq} D^q s_x$$

where $X \subset R^n$, X finite, nonvoid, $p_x \in N^n$ and $\lambda_{xq} \in C^1$. Moreover

(10) $$\exists \; x_o \in X \;, \; q_o \in N^n \;, \; q_o \leq p_{x_o} \;:\; \lambda_{x_o q_o} \neq 0$$

We show that (17.2) is satisfied for x_o given in (10). Assume it is false and

(11) $$\exists \; \mu \in N \;:\; \forall \; \nu \in N \;, \; \nu \geq \mu \;:\; t(\nu)(x_o) = 0$$

then (9) gives

(12) $$\sum_{x \in X} \sum_{\substack{q \in N^n \\ q \leq p_x}} \lambda_{xq} D^q s_x(\nu)(x_o) = 0 \;, \quad \forall \; \nu \in N \;, \; \nu \geq \mu$$

But, due to (5) and the fact that X is finite, one can take μ such that (12) implies

(13) $$\sum_{\substack{q \in N^n \\ q \leq p_{x_o}}} \lambda_{x_o q} D^q s_{x_o}(\nu)(x_o) = 0 \;, \quad \forall \; \nu \in N \;, \; \nu \geq \mu$$

Define now the infinite vector of complex numbers $\Lambda = (\lambda'_\mu \mid \mu \in N)$ where

(14) $$\lambda'_\mu = \begin{cases} \lambda_{x_o p(\mu+1)} & \text{if } p(\mu+1) \leq p_{x_o} \\ 0 & \text{otherwise} \end{cases}$$

then, obviously $\Lambda \in M$ and (13) is equivalent to

$$W(s_{x_o})(x_o)\ \Lambda \in M$$

Therefore, (6) will imply $\Lambda = 0$ which through (14) will contradict (10), ending the proof of (17.2) in chap. 2.

It remains to prove that T_Σ and I^δ are compatible, that is, the relations (4) and (5) in chap. 2, §3, are satisfied. The relation (5) follows easily since I^δ is a Dirac ideal. In order to prove (4) it suffices to show that

(15) $$I^\delta \cap T_\Sigma = 0$$

since $V_o \cap T_\Sigma = 0$, as it was noticed above. Assume therefore $t \in I^\delta \cap T_\Sigma$, $t \notin 0$, then (9) and (10) hold again, since $t \in T_\Sigma$. But, $t \in I^\delta$ and (2) will again give (11) thus the reasoning above, contradicting (10) will end the proof of (15) ▽▽▽

§4. PRODUCTS WITH DIRAC DISTRIBUTIONS

Based on the compatibility of the Dirac ideal I^δ with the Dirac classes T_Σ , for $\Sigma \in Z_\delta$, we shall follow the procedure in Theorem 1, chap. 2, §3, and construct the Dirac algebras used in the present chapter.

Suppose given $\Sigma \in Z_\delta$.
For any ideal I in W , $I \supset I^\delta$ and compatible vector subspace T in S_o , $T \supset T_\Sigma$, if V is a vector subspace in $I \cap V_o$ and S_1 is a vector subspace in S_o such that

(16) $$V_o \oplus T \oplus S_1 = S_o ,$$

(17) $$U \subset V(p) \oplus T \oplus S_1 , \quad \forall\ p \in \bar{N}^n ,$$

then $(V, T \oplus S_1)$ is a regularization, therefore one can define for any admissible property Q the Dirac algebras

(18) $$A^Q(V, T \oplus S_1 , p) , \quad p \in \bar{N}^n$$

which for the sake of simplicity will be denoted within the present chapter by A_p , with $p \in \bar{N}^n$.

Properties of type (1) concerning products with Dirac distributions are given in:

Theorem 1

In case

(19) $$I^\delta \cap V_o \subset V$$

any q-th order $(q \in N^n)$ derivative $D^q\delta_{x_o}$ of the Dirac delta distribution in $x_o \in R^n$ has the properties:

1) $$D^q\delta_{x_o} \neq 0 \in A_p , \quad \forall\ p \in \bar{N}^n$$

2) $\psi(x-x_o) \cdot D^q\delta_{x_o} = 0 \in A_p \,, \quad \forall \; p \in \bar{N}^n \,,$

for every $\psi \in C^\infty(R^n)$ which satisfies

(20) $D^r\psi(0) = 0 \,, \quad \forall \; r \in N^n \,, \quad r \leq p \;$ or $\; r \leq q$

in particular

3) $(x-x_o)^r \cdot D^q\delta_{x_o} = 0 \in A_p \,, \quad \forall \; p \in N^n \,, \quad r \in N^n \,, \quad r \nleq p$ and $r \nleq q$

Proof

1) Assume $\Sigma = (s_x \mid x \in R^n)$. According to 3) in Theorem 2, chap. 1, §8, the relation holds

$$D^q\delta_{x_o} = D^q s_{x_o} + I^Q(V(p) \,, \, T \oplus S_1) \in A_p$$

therefore, $D^q\delta_{x_o} = 0 \in A_p$, only if $D^q s_{x_o} \in I^Q(V(p), T \oplus S_1)$. But, obviously $D^q s_{x_o} \in T_\Sigma$. Thus $D^q\delta_{x_o} = 0 \in A_p$ implies $D^q s_{x_o} \in T_\Sigma \cap I^Q(V(p), T \oplus S_1) \subset T \cap I = 0$ and the relation $D^q s_{x_o} \in 0$ is absurd due to (4).

2) According again to 3) in Theorem 2, chap. 1, §8, the relation holds

(21) $$\psi(x-x_o) \cdot D^q\delta_{x_o} = \psi(x-x_o) \cdot D^q s_{x_o} + I^Q(V(p), T \oplus S_1) \in A_p$$

Define $v \in W$ by

$$v(\nu)(x) = \psi(x-x_o) \cdot D^q s_{x_o}(\nu)(x) \,, \quad \forall \; \nu \in N \,, \quad x \in R^n \,,$$

then (21) becomes

(22) $$\psi(x-x_o) \cdot D^q\delta_{x_o} = v + I^Q(V(p), T \oplus S_1) \in A_p$$

We shall prove that

(23) $$v \in I^Q(V(p), T \oplus S_1)$$

and then, due to (22), the proof of 2) in Theorem 1 will be completed. First we notice that $v \in S_o$ since $\psi \in C^\infty(R^n)$ and (4). Actually, $v \in V_o$ since $r \leq q$ in (20). Thus

(24) $$D^r v \in V_o \,, \quad \forall \; r \in N^n \,.$$

Assume now $r \in N^n$, $r \leq p$, then (20) and (5) will imply that $D^r v \in I^\delta$ which together with (24) and (19) will give

$$D^r v \in I^\delta \cap V_o \subset V \,, \quad \forall \; r \in N^n \,, \quad r \leq p$$

That relation implies $v \in V(p)$ and the proof of (23) is completed.

3) It results from 2) choosing $\psi(x) = x^r$, $\forall \; x \in R^n$ ∇∇∇

Remark 1

The relations in 3) in Theorem 1, describing the products within the algebras containing the distributions between polynomials and derivatives of the Dirac δ distribution are identical with the usual formulas (1) within $\mathcal{D}'(R^n)$, except when $r \leq p$. However, even in that case, the relations proved in the algebras are also valid in $\mathcal{D}'(R^n)$. For instance, in the one dimensional case $n = 1$, the relations in 3) in Theorem 1, imply for any $p \in N$ and $x_0 \in R^1$:

$$(25) \qquad (x-x_0)^{p+1} \cdot \delta_{x_0} = (x-x_0)^{p+1} \cdot D\delta_{x_0} = \ldots = (x-x_0)^{p+1} \cdot D^{p-1}\delta_{x_0} = 0 \in A_p ,$$

$$(26) \qquad (x-x_0)^{q+1} \cdot D^q\delta_{x_0} = 0 \in A_p , \quad \forall\ q \in N , \ q \geq p$$

The nontriviality of the powers of Dirac distribution derivatives is given in:

Theorem 2

In case

$$(27) \qquad V \subset \mathcal{I}^{\delta} \cap V_0$$

the relations hold

$$(D^q\delta_{x_0})^k \neq 0 \in A_p , \quad \forall\ p \in \bar{N}^n , \ x_0 \in R^n , \ q \in N^n , \ k \in N , \ k \geq 1$$

Proof

Assume $\Sigma = (s_x \mid x \in R^n)$. According to 3) in Theorem 2, chap. 1, §8, the relation holds

$$(D^q\delta_{x_0})^k = (D^q s_{x_0})^k + I^Q(V(p),\mathcal{T} \oplus \mathcal{S}_1) \in A_p$$

therefore, defining $v \in W$ by $v = (D^q s_{x_0})^k$, the theorem is valid, only if $v \notin I^Q(V(p),\mathcal{T} \oplus \mathcal{S}_1)$.

Assume, it is false, then (27) and (2) imply for a certain $\mu \in N$ the relation

$$v(\nu)(x_0) = 0 , \quad \forall\ \nu \in N , \ \nu \geq \mu$$

which due to the definition of v , will result in

$$D^q s_{x_0}(\nu)(x_0) = 0 , \quad \forall\ \nu \in N , \ \nu \geq \mu .$$

However, that relation obviously contradicts (6) ∇∇∇

The nontriviality of the product of Dirac distribution derivatives concentrated in the same point of R^n can be obtained in special cases:

Theorem 3

In case (27) in Theorem 2 is valid, there exist $\Sigma \in Z_\delta$ such that in the corresponding algebras A_p , $p \in \bar{N}^n$, the relations hold

$$D^{q_0}\delta_{x_0} \cdot \ldots \cdot D^{q_k}\delta_{x_0} \neq 0 \in A_p , \quad \forall \ p \in \bar{N}^n , x_0 \in R^n , k \in N ,$$
$$q_0 , \ldots , q_k \in N^n$$

Proof

Assume $\Sigma = (s_x \mid x \in R^n)$ with s_x given in the proof of Corollary 1, §7. According to 3) in Theorem 2, chap. 1, §8, the relation holds

$$(28) \qquad D^{q_0}\delta_{x_0} \cdot \ldots \cdot D^{q_k}\delta_{x_0} = D^{q_0}s_{x_0} \cdot \ldots \cdot D^{q_k}s_{x_0} + I^Q(V(p),T + S_1) \in A_p$$

Define $v \in W$ by

$$(29) \qquad v = D^{q_0}s_{x_0} \cdot \ldots \cdot D^{q_k}s_{x_0}$$

then, due to (28), the theorem holds only if $v \notin I^Q(V(p),T \oplus S_1)$. Assume $v \in I^Q(V(p),T \oplus S_1)$, then (27) and (2) imply for a certain $\mu \in N$ the relation

$$v(\nu)(x_0) = 0 , \quad \forall \ \nu \in N , \ \nu \geq \mu$$

Now, (29) gives

$$D^{q_0}s_{x_0}(\nu)(x_0) \cdot \ldots \cdot D^{q_k}s_{x_0}(\nu)(x_0) = 0 , \quad \forall \ \nu \in N , \ \nu \geq \mu$$

which implies the existence of $0 \leq i \leq k$ such that $D^{q_i}s_{x_0}(\nu)(x_0)$ vanishes for infinitely many values of $\nu \in N$.
However, that contradicts the fact that $D^p\psi(0) = 1 / K > 0$, $\forall \ p \in N^n$, established in the proof of Corollary 1, §7 $\nabla\nabla\nabla$

An expected property of the product of two Dirac distribution derivatives concentrated in different points in R^n , is given in:

Theorem 4

In case (19) in Theorem 1 is valid, one obtains the relations

$$D^q\delta_x \cdot D^r\delta_y = 0 \in A_p , \quad \forall \ p \in \bar{N}^n , \ x,y \in R^n , \ x \neq y , \ q,r \in N^n$$

Proof

Assume $\Sigma = (s_x \mid x \in R^n)$. According to 3) in Theorem 2, chap. 1, §8, the relation holds

$$(30) \qquad D^q\delta_x \cdot D^r\delta_y = D^q s_x \cdot D^r s_y + I^Q(V(p),T \oplus S_1) \in A_p$$

Denoting $v = D^q s_x \cdot D^r s_y$, the relation (5) together with $x \neq y$ implies $D^h v \in I^\delta \cap V_0$, $\forall \ h \in N^n$. Therefore $v \in V(p) \subset I^Q(V(p),T \oplus S_1)$ and the relation (30) will end the proof $\nabla\nabla\nabla$

In the case of derivative algebras, the relations in 3) in Theorem 1 - see also (25) and (26) in Remark 1 - will be supplemented in Theorem 5. For the sake of simplicity, the one dimensional case $n = 1$ is considered only.

Theorem 5

In case (19) in Theorem 1 is valid, one obtains the relations

$$(x-x_0)^q \cdot (D^q \delta_{x_0})^k = 0 \in A_p , \quad \forall \; p \in N , \; x_0 \in R^1 , \; q \in N , \; q \geq p+1 , \; k \in N , \; k \geq 2$$

Proof

Assume $\Sigma = (s_x \mid x \in R^1)$. According to 1) and 4) in Theorem 3, chap. 1, §8, the relation

(31)
$$D^1_{p+1}((x-x_0)^{q+1} \cdot (D^r \delta_{x_0})^k) = (q+1)(x-x_0)^q \cdot (D^r \delta_{x_0})^k + (x-x_0)^{q+1} \cdot k \cdot (D^r \delta_{x_0})^{k-1} \cdot D^{r+1} \delta_{x_0}$$

holds in A_p , for any $q,r,k \in N$. Now, due to 3) in Theorem 1, one obtains

(32)
$$(x-x_0)^{q+1} \cdot D^r \delta_{x_0} = 0 \in A_p , \quad \forall \; q,r \in N , \; q \geq p , \; q \geq r$$

Therefore, (31) and (32) imply in A_p the relation

(33)
$$D^1_{p+1}((x-x_0)^{q+1} \cdot (D^r \delta_{x_0})^k) = (q+1(x-x_0)^q \cdot (D^r \delta_{x_0})^k , \quad \forall \; q,r,k \in N , \; q \geq p , \; q \geq r , \; k \geq 2$$

But, the product in the left side of (33) is computed in A_{p+1} and according to 3) in Theorem 1, one obtains

(34)
$$(x-x_0)^{q+1} \cdot D^r \delta_{x_0} = 0 \in A_{p+1} , \quad \forall \; q,r \in N , \; q \geq p+1 , \; q \geq r$$

Taking $r = q$, the relations (33) and (34) will end the proof ∇∇∇

An example for the application of Theorem 1 is given in:

Proposition 3

The Riccati differential equation

$$y' = x^{q+r} \cdot y \cdot (y+1) + D^{r+1} \delta(x) , \text{ with } x \in R^1 , \; q,r \in N , \; q \geq 1 ,$$

has in the algebras A_p , $p \leq r$, the general solution

$$y(x) = 1 / (c \exp(-x^{q+r+1} / (q+r+1))-1) + D^r \delta(x) , \quad x \in R^1 , \; c \in (-\infty,0] ,$$

provided that (19) is fulfiled.

Proof

Assume $c \in R^1$ and define $\psi \in C^\infty(R^1)$ by

$$\psi(x) = c \exp(-x^{q+r+1}/(q+r+1)) - 1 , \quad x \in R^1 ,$$

then $1/\psi \in C^\infty(R^1)$, for $c \in (-\infty,0]$. Therefore

$$T = 1/\psi + D^r\delta \in D'(R^1) \subset A_p , \quad \forall \ p \in \bar{N} , \ c \in (-\infty,0] .$$

Since A_p , with $p \in \bar{N}$, are associative, commutative and with the unit element $1 \in C^\infty(R^1)$, one obtains in these algebras the relation

$$x^{q+r} \cdot T \cdot (T+1) = x^{q+r} \cdot (D^r\delta)^2 + 2\, x^{q+r} \cdot (D^r\delta) \cdot (1/\psi) +$$

$$+ x^{q+r} \cdot (1/\psi)^2 + x^{q+r} \cdot (D^r\delta) + x^{q+r} \cdot (1/\psi)$$

provided that $c \in (-\infty,0]$. But, due to 3) in Theorem 1, one obtains in A_p , with $p \leq r$, the relation

$$x^{q+r} \cdot D^r\delta = 0 \in A_p$$

since $q + r > r$, as $q \geq 1$. Therefore, one obtains in the algebras A_p , with $p \leq r$, the relation

$$x^{q+r} \cdot T \cdot (T+1) = x^{q+r} \cdot (1/\psi) \cdot (1/\psi+1) , \quad \forall \ c \in (-\infty,0] ,$$

which means that T is a solution of the considered Riccati equation ∇∇∇

§5. FORMULAS IN QUANTUM MECHANICS

In the one dimensional case $n = 1$, the Dirac δ distribution and the Heisenberg distributions

$$\delta_+ = (\delta + (1/x) / \pi i) / 2$$

$$\delta_- = (\delta - (1/x) / \pi i) / 2$$

satisfy the formulas, [108], given in:

Theorem 6

There exist $\Sigma \in Z_\delta$ and regularizations $(V, T \oplus S_1)$ (see the beginning of §4) such that within the corresponding algebras A_p , $p \in \bar{N}$, the relations are valid:

(35) $$(\delta)^2 - (1/x)^2 / \pi^2 = -(1/x^2) / \pi^2$$

(36) $$(\delta_+)^2 = -D\delta / 4\pi i - (1/x^2) / 4\pi^2$$

(37) $$(\delta_-)^2 = D\delta / 4\pi i - (1/x^2) / 4\pi^2$$

(38) $$\delta \cdot (1/x) = -D\delta / 2$$

Proof

Assume $\psi \in \mathcal{D}(R^1)$ with $\int_{R^1} \psi(x)dx = 1$ and satisfying (8), then s_ψ given by (7) will belong to Z_o.
Denoting $M_q = \sup \{ \, |D^q\psi(x)| \; | \; x \in R^1 \}$, for $q \in N$, and assuming that $\text{supp}\,\psi \subset [-L,L]$, for a certain $L > 0$, one obtains

$$| \, x^{q+1} \cdot D^q s_\psi(\nu)(x) \, | \leq M_q \cdot L^q \, , \quad \forall \; q,\nu \in N \, , \quad x \in R^1 \, ,$$

therefore, s_ψ is a 'δ sequence' according to [106], [108].
Assume $\Sigma = (s_x \mid x \in R^1) \in Z_\delta$ such that $s_o = s_\psi$. Define the sequence of smooth functions $t \in W$ by the convolutions $t(\nu) = s_\psi(\nu) * (1/x)$, with $\nu \in N$. Then, obviously

$$t \in S_o \quad \text{and} \quad < t \, , \cdot > = 1/x$$

Denote by P_t the vector subspace in S_o generated by $\{ \, D^q t \mid q \in N \}$, then

$$(V_o \oplus U \oplus T_\Sigma) \cap P_t = O$$

according to Lemmas 1 and 2, below. Therefore, one can choose a vector subspace S_1 in S_o which satisfies

(39) $$V_o \oplus T_\Sigma \oplus S_1 = S_o$$

(40) $$U \oplus P_t \subset S_1$$

Taking now $T = T_\Sigma$, the relations (39) and (40) will imply (16) and (17), therefore $(V, T + S_1)$ will be a regularization.

Since $s_\psi \in T_\Sigma = T$ and $t \in P_t \subset S_1$, one obtains according to 3) in Theorem 2, chap. 1, §8, the relations

(41) $$\delta_+ = (s_\psi + t/\pi i) \, / \, 2 + I^Q(V(p), T \oplus S_1) \in A_p$$

(42) $$\delta_- = (s_\psi - t/\pi i) \, / \, 2 + I^Q(V(p), T \oplus S_1) \in A_p$$

(43) $$\delta \cdot (1/x) = s_\psi \cdot t + I^Q(V(p), T \oplus S_1) \in A_p$$

Define the sequences of smooth functions t_1, t_2, $t_3 \in W$ by

(44) $$t_1 = (s_\psi + t/\pi i)^2 \, , \quad t_2 = (s_\psi - t/\pi i)^2 \, , \quad t_3 = s_\psi \cdot t$$

It was proved in [108] that t_1, t_2, $t_3 \in S_o$ and

(45) $$< t_1 \, , \cdot > = -D\delta \, / \, \pi i - (1/x^2) \, / \, \pi^2$$

(46) $$< t_2 \, , \cdot > = D\delta \, / \, \pi i - (1/x^2) \, / \, \pi^2$$

(47) $$< t_3 \quad \cdot > = -D\delta \, / \, 2$$

The relations (41-43) and (45-47) will give through (44), the required relations (36-38). It only remains to prove (35). From the definition of δ_+ it follows that in each algebra A_p, with $p \in \bar{N}$, the relation holds

$$(\delta_+)^2 = (\delta + (1/x)/\pi i)^2 \, / \, 4 = (\delta)^2 \, / \, 4 - (1/x)^2 \, / \, 4\pi^2 + \delta \cdot (1/x) \, / \, 2\pi i$$

which compared with (36) and (38) will give (35) ∇∇∇

And now, the two lemmas concerning distributions in $\mathcal{D}'(R^1)$ needed in the proof of Theorem 6.

Denote by $\mathcal{D}'_\delta(R^1)$ the set of all distributions in $\mathcal{D}'(R^1)$ with support a finite subset of R^1. Denote by S_δ the set of all weakly convergent sequences of smooth functions $s \in S_0$ which generate distributions $<s,\cdot>$ in $\mathcal{D}'_\delta(R^1)$. Finally, denote by $\mathcal{D}'_\sigma(R^1)$ the set of all distributions $T \in \mathcal{D}'(R^1)$ such that $\sum_{0 \le r \le q} \lambda_r D^r T \in \mathcal{D}'_\delta(R^1)$ for certain $q \in N$, $\lambda_r \in C^1$, $\lambda_q \neq 0$.

<u>Lemma 1</u>

For $t \in S_0$ denote by P_t the vector subspace in S_0 generated by $\{ D^q t \mid q \in N \}$. If $t \notin \mathcal{O}$, then $(\mathcal{U} \oplus S_\delta) \cap P_t = \mathcal{O} \Longleftrightarrow <t,\cdot> \notin C^\infty(R^1) + \mathcal{D}'_\sigma(R^1)$

<u>Proof</u>

The implication $\Leftarrow$. Assume, it is false and let $s \in (\mathcal{U} \oplus S_\delta) \cap P_t$ be such that $t \notin \mathcal{O}$. Then

$$(48) \qquad s = u(\psi) + t_1 = \sum_{0 \le r \le q} \lambda_r D^r t$$

for certain $\psi \in C^\infty(R^1)$, $t_1 \in S_\delta$, $q \in N$, $\lambda_r \in C^1$, $\lambda_q \neq 0$.
Denote $P(D) = \sum_{0 \le r \le q} \lambda_r D^r$. Let $\chi \in C^\infty(R^1)$ be such that $P(D)\chi = \psi$. Then

$$(49) \qquad t_2 = t - u(\chi) \in S_0 , \quad < t_2 , \cdot > \in \mathcal{D}'(R^1)$$

since $P(D)t_2 = t_1$, due to (48), while $t_1 \in S_\delta$. But (49) implies $<t,\cdot> = \chi + <t_2, \cdot> \in C^\infty(R^1) + \mathcal{D}'_\sigma(R^1)$ contradicting the hypothesis.

Now, the implication $\Rightarrow$. Assume, it is false and $<t,\cdot> \in C^\infty(R^1) + \mathcal{D}'_\sigma(R^1)$. Then

$$< t , \cdot > = \psi + < t_1 , \cdot >$$

with $\psi \in C^\infty(R^1)$ and $t_1 \in S_0$ such that $s = P(D)t_1 \in S_\delta$. Therefore, $<P(D)t,\cdot> = P(D)\psi + <s,\cdot>$, hence $P(D)t = u(\psi) + s + v$, for certain $v \in V_0$. It follows that $P(D)t \in \mathcal{U} \oplus S_\delta$. Now, if $P(D)t \notin \mathcal{O}$ then $(\mathcal{U} \oplus S_\delta) \cap P_t \ni P(D)t \notin \mathcal{O}$ contradicting the hypothesis. On the other side, if $P(D)t \in \mathcal{O}$ then $P(D) <t,\cdot> = <P(D)t,\cdot> = 0 \in \mathcal{D}'(R^1)$, hence $<t,\cdot> \in C^\infty(R^1)$, therefore $t \in \mathcal{U} \oplus S_\delta$, since $V_0 \subset S_\delta$. One obtains finally $(\mathcal{U} \oplus S_\delta) \cap P_t \ni t \notin \mathcal{O}$ again contradicting the hypothesis ∇∇∇

<u>Lemma 2</u>

$(1/x^m) \notin C^\infty(R^1) + \mathcal{D}'_\sigma(R^1)$, $\forall\ m \in N$, $m \ge 1$.

Proof

Assume, it is false. Then

(50) $$1/x^m = \psi + T$$

for certain $\psi \in C^\infty(R^1)$ and $T \in D'_\sigma(R^1)$. Hence

(51) $$T \Big|_{R^1\setminus\{0\}} = \chi$$

for certain $\chi \in C^\infty(R^1\setminus\{0\})$. But, according to the definition of $D'_\sigma(R^1)$, it follows that

(52) $$S = \sum_{0\leq r\leq q} \lambda_r D^r T \in D'_\delta(R^1)$$

for certain $q \in N$, $\lambda_r \in C^1$, $\lambda_q \neq 0$. Then.

(53) $$S_1 = S \Big|_{R^1\setminus\{0\}} \in D'_\delta(R^1\setminus\{0\})$$

Now, (51-53) imply

(54) $$S_1 = P(D)\chi \in C^\infty(R^1\setminus\{0\})$$

where $P(D) = \sum_{0\leq r\leq q} \lambda_r D^r$. As $C^\infty \cap D'_\delta = \{0\}$, the relations (53), (54) result in $S_1 = 0$ which together with (50-52) gives

$$P(D)(1/x^m) = P(D)\psi \quad \text{on} \quad R^1\setminus\{0\}$$

Computing the derivative in the left side, one obtains

$$\sum_{0\leq r\leq q} (-1)^r \frac{(m+r-1)!}{(m-1)!} \lambda_r x^{q-r} = x^{m+q} P(D)\psi(x) , \quad \forall x \in R^1\setminus\{0\}$$

Taking the limit for $x \to 0$, one obtains

$$(-1)^q \frac{(m+q-1)!}{(m-1)!} \lambda_q = 0$$

contradicting the assumption that $\lambda_q \neq 0$ ∇∇∇

§6. A PROPERTY OF THE DERIVATIVE IN THE ALGEBRAS

In the present section, the case of derivative algebras containing $D'(R^1)$ will be considered.

According to the general result in 1) in Theorem 3, chap. 1, §8, the derivative mappings within the algebras

$$D^q_{p+q} : A_{p+q} \to A_p , \quad p \in \bar{N} , \quad q \in N ,$$

coincide on $C^\infty(R^1)$ with the usual derivatives D^q of smooth functions. That result will be strengthened in Theorem 7.

First, we notice that due to the inclusion $\mathcal{T}_{\Sigma} \subset \mathcal{T}$, the same 1) in Theorem 3, chap. 1, §8, implies that the derivative mappings within the algebras coincide on $\mathcal{C}^{\infty}(R^1) \oplus \mathcal{D}'_{\delta}(R^1)$ with the usual distribution derivatives.

Theorem 7

Given any distribution $T \in \mathcal{D}'(R^1) \setminus (\mathcal{C}^{\infty}(R^1) + \mathcal{D}'_{\sigma}(R^1))$ there exist regularizations $(\mathcal{V}, \mathcal{T} \oplus \mathcal{S}_1)$ such that within the corresponding algebras A_p , $p \in \bar{N}$, the derivative mappings coincide on $\mathcal{C}^{\infty}(R^1) + \mathcal{D}'_{\sigma}(R^1) + \mathcal{M}_T$ with the usual distribution derivatives, where $\mathcal{M}_T$ is the vector subspace in $\mathcal{D}'(R^1)$ generated by $\{ T , DT , D^2T , \ldots \}$.

Proof

Assume $T = \langle t, \cdot \rangle$ for a certain $t \in \mathcal{S}_0$. Then, according to Lemma 1, §5, $(\mathcal{U} \oplus \mathcal{S}_{\delta}) \cap \mathcal{P}_t = \mathcal{O}$. But, obviously $\mathcal{S}_{\delta} = \mathcal{V}_0 \oplus \mathcal{T}_{\Sigma}$, for any $\Sigma \in \mathcal{Z}_{\delta}$. Therefore, given $\Sigma \in \mathcal{Z}_{\delta}$, one can choose a vector subspace $\mathcal{S}_1$ in $\mathcal{S}_0$ such that

(55) $$\mathcal{V}_0 \oplus \mathcal{T}_{\Sigma} \oplus \mathcal{S}_1 = \mathcal{S}_0$$

(56) $$\mathcal{U} \oplus \mathcal{P}_t \subset \mathcal{S}_1$$

Taking $\mathcal{T} = \mathcal{T}_{\Sigma}$, the relations (55), (56) will imply (16) and (17), therefore $(\mathcal{V}, \mathcal{T} \oplus \mathcal{S}_1)$ will be a regularization. Noticing that

$$\mathcal{M}_T = \{ \langle s, \cdot \rangle \mid s \in \mathcal{P}_t \}$$

and taking into account 1) in Theorem 3, chap. 1, §8, the proof is completed ∇∇∇

§7. THE EXISTENCE OF THE SEQUENCES IN Z_0

In order to prove that (see §3)

$$\mathcal{Z}_{x_0} \neq \emptyset , \quad \forall \ x_0 \in R^n ,$$

it is obviously sufficient to show that $\mathcal{Z}_0 \neq \emptyset$. In this respect, a class of sequences s belonging to $\mathcal{Z}_0$ will be constructed by a proper generalization to $n \geq 1$ dimensions of the method in (7) and (8).

Suppose $\psi \in \mathcal{D}(R^n)$ such that $\int_{R^n} \psi(x)dx = 1$ and define $s_{\psi} \in \mathcal{W}$ by

(57) $$s_{\psi}(\nu)(x) = \mu_1(\nu) \cdot \ldots \cdot \mu_n(\nu) \cdot \psi(\mu_1(\nu)x_1 , \ldots, \mu_n(\nu)x_n) ,$$
$$\forall \ \nu \in N , \ x = (x_1 , \ldots, x_n) \in R^n$$

where the mapping

(58) $$N \ni \nu \to \mu(\nu) = (\mu_1(\nu), \ldots, \mu_n(\nu)) \in N^n$$

is constructed in (59-62).

First, define $N \ni \nu \to k(\nu) \in N$, by (see §3):

(59) $$k(0) = 0 \quad \text{and} \quad k(\nu+1) = k(\nu) + \ell(n,\nu+1) \ , \quad \forall \ \nu \in N \ .$$

Define also $N \ni \nu \to h(\nu) \in N$, by

(60) $$h(0) = 0 \quad \text{and} \quad h(\nu+1) = h(\nu) + \nu + 1 \ , \quad \forall \ \nu \in N \ .$$

Define now $N \ni \nu \to e(\nu) \in N^n$, by

(61) $$e(\nu) = (h(\nu),\ldots,h(\nu))) \ , \quad \forall \ \nu \in N \ .$$

Finally, define (58), by

(62.1) $$\mu(0) = (1,\ldots,1) \in N^n$$

(62.2) $$\{ \mu(k(\nu)+1) \ , \ \ldots \ , \ \mu(k(\nu+1)) \} = P(n,\nu+1) + e(\nu+1) \ , \quad \forall \ \nu \in N \ .$$

The mapping (58) is illustrated in Fig. 4, in the case of $n = 2$. There, the set denoted by M_4 can be written in terms of (62.2), as

$$M_4 = \{ \mu(k(3)+1) \ , \ \ldots \ , \ \mu(k(4)) \} = P(2,4) + e(4)$$

<u>Lemma 3</u>

s_ψ satisfies (4) and (5).

<u>Proof</u>

It follows from the fact that

$$\lim_{\nu\to\infty} \mu_i(\nu) = +\infty \ , \quad \forall \ 1 \le i \le n \quad \nabla\nabla\nabla$$

The basic property of the sequences s_ψ defined in (57-62) is given in:

<u>Proposition 4</u>

The following three conditions are equivalent:

*) $s_\psi \in Z_0$

**) $W(s_\psi)(0)$ is column wise nonsingular

***) $D^p\psi(0) \neq 0$, $\forall \ p \in N^n$ (see(8))

<u>Proof</u>

Taking into account the definition of Z_0 in §3 as well as Lemma 3, the conditions *) and **) are obviously equivalent. It only remains to establish the equivalence between **) and ***). First, we compute $W(s_\psi)(0)$. The relation (57) will give easily

$$D^q s_\psi(\nu)(0) = (\mu(\nu))^{q+e} D^q\psi(0) \ , \quad \forall \ q \in N^n \ , \ \nu \in N \ ,$$

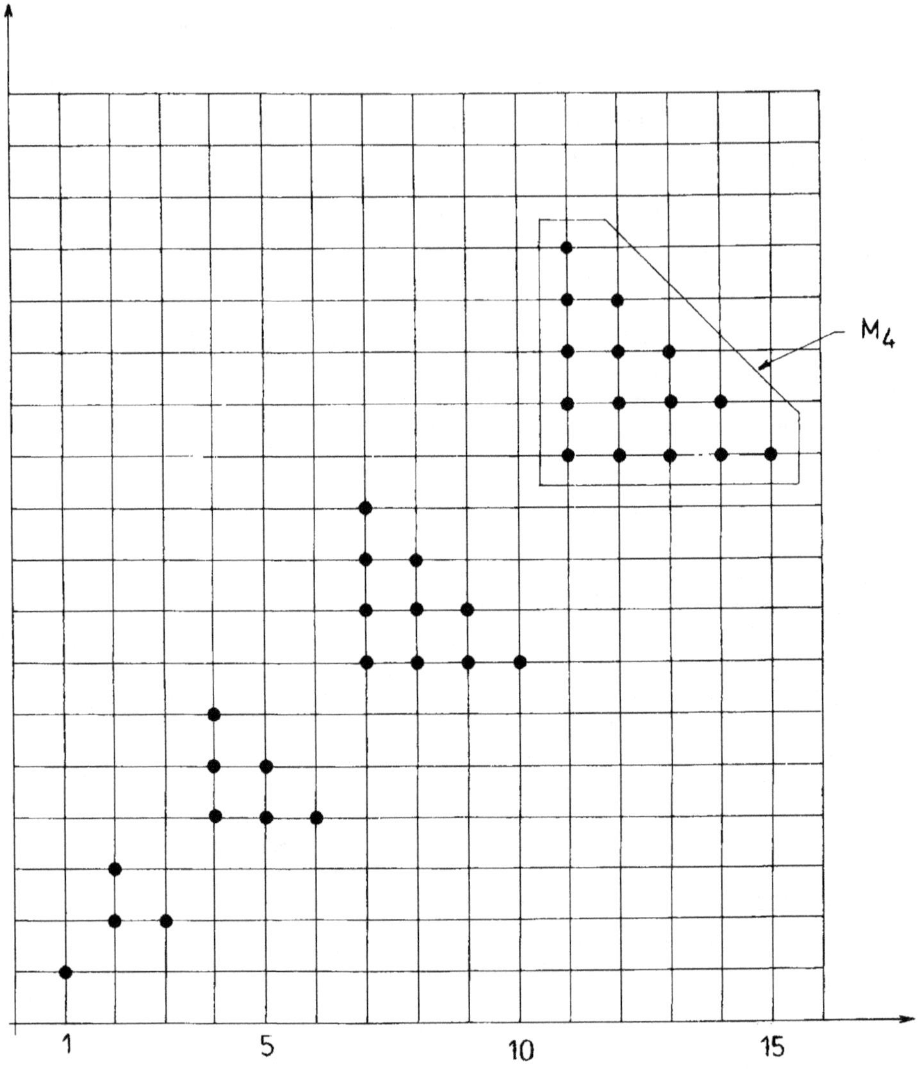

Fig. 4

where $e = (1,\ldots,1) \in N^n$. Therefore

(63) $$W(s_\psi)(0) = A \cdot B$$

where

$$A = ((\mu(\nu))^{p(\sigma)+e} \mid \nu,\sigma \in N)$$

while B is a diagonal matrix with the diagonal elements

(64) $$D^{p(\nu)}\psi(0) , \quad \nu \in N$$

Now, according to Theorem 8, below, A is columnwise nonsingular. Indeed, due to Lemma 4, below, it suffices to show that A satisfies (65). Assume given $\bar{\nu},\bar{\sigma} \in N$. We choose $m \in N$, such that $\sigma = \ell(n,m+1) - 1 \geq \bar{\sigma}$ and $k(m) \geq \bar{\nu}$. Now, we choose

$$\nu_0 = k(m) + 1 ,\ldots, \nu_\sigma = k(m+1)$$

Then, the conditions *) and **) in (65) are obviously satisfied, while (62.2) and Theorem 8, will directly imply ***) in (65). Therefore A is column wise nonsingular. Now, the relations (63), (64) and Lemma 4, imply that $W(s_\psi)(0)$ is column wise nonsingular, only if $D^{p(\nu)}\psi(0) \neq 0$, $\forall \, \nu \in N$ ▽▽▽

And now, the main result of the present section

<u>Corollary 1</u>

$$Z_{x_0} \neq \emptyset , \quad \forall \; x_0 \in R^n$$

<u>Proof</u>

According to Proposition 4, it suffices to show the existence of $\psi \in \mathcal{D}(R^n)$ such that $\int_{R^n} \psi(x)dx = 1$ and $D^p\psi(0) \neq 0$, $\forall \; p \in N^n$.

Define $\alpha : R^n \to R^1$ by $\alpha(x_1 , \ldots ,x_n) = \exp(x_1 +\ldots+ x_n)$ and assume $\beta \in \mathcal{D}(R^n)$ such that $\beta \geq 0$ and $\beta = 1$ in a certain neighbourhood of $0 \in R^n$. Then

$$K = \int_{R^n} \alpha(x)\beta(x)dx > 0$$

Defining $\psi = \alpha \cdot \beta / K$, one obtains the required function, since $D^p\psi(0) = 1/K > 0$ $\forall \; p \in N^n$ ▽▽▽

In case arbitrary positive powers of the Dirac δ distributions are to be defined within the algebras constructed in the present chapter (see Theorem 4, chap. 1, §8 and chap. 4) one needs the result given in:

<u>Corollary 2</u>

$$Z_{x_0} \cap \mathcal{W}_+ \neq \emptyset , \quad \forall \; x_0 \in R^n$$

Proof

Choosing $\beta \in C_+^\infty(R^n)$ in the proof of Corollary 1, one obtains $\psi \in C_+^\infty(R^n)$ and therefore $s_\psi \in Z_o \cap W_+$ ▽▽▽

Lemma 4

The infinite matrix of complex numbers $A = (a_{\nu\sigma} \mid \nu,\sigma \in N)$ is column wise non-singular, only if

$\forall\ \bar{\nu},\bar{\sigma} \in N :$

$\exists\ \sigma \in N\ ,\ \nu_o ,\dots, \nu_\sigma \in N :$

*) $\bar{\sigma} \le \sigma$

**) $\bar{\nu} \le \nu_o < \dots < \nu_\sigma$

***)
$$\text{(65)}\qquad \begin{vmatrix} a_{\nu_o o} & \cdots & a_{\nu_o \sigma} \\ \vdots & & \vdots \\ a_{\nu_\sigma o} & \cdots & a_{\nu_\sigma \sigma} \end{vmatrix} \neq 0$$

Proof

It follows easily from the definition in §3 ▽▽▽

And now, the theorem on generalized Vandermonde determinants (for notations, see §3), whose present form, as well as proof was offered by R.C. King.

Theorem 8

Suppose given $n \in N$, $n \ge 1$.
Then, for each $a \in N^n$, $a \ge e = (1,\dots,1) \in N^n$ and $\ell \in N$, $\ell \ge 1$, the relation holds

$$\begin{vmatrix} (a+p(1))^{p(1)} & \cdots & (a+p(1))^{p(\ell)} \\ \vdots & & \vdots \\ (a+p(\ell))^{p(1)} & \cdots & (a+p(\ell))^{p(\ell)} \end{vmatrix} = \prod_{1\le i\le n}\ \prod_{1\le j\le \ell} (p_i(j))! > 0$$

where $p(j) = (p_1(j),\dots,p_n(j))$, for $1 \le j \le \ell$.

Remark

The value of the determinant depends only on $n,\ell,p(1),\dots,p(\ell)$ and does not depend on a .

Proof

Let us consider the determinant

$$\Delta_1 = \det\left((a+p(\sigma))^{p(\tau)}\right), \text{ where } 1 \le \sigma, \tau \le \ell$$

For $1 \le \tau \le \ell$, the τ-th column in Δ_1 is

$$C_1(\tau) = \left|\begin{matrix} (a_1+p_1(1))^{p_1(\tau)} \times \dots \times (a_n+p_n(1))^{p_n(\tau)} \\ \vdots \\ (a_1+p_1(\ell))^{p_1(\tau)} \times \dots \times (a_n+p_n(\ell))^{p_n(\tau)} \end{matrix}\right|$$

if $a = (a_1, \dots, a_n)$.

For $1\le\tau\le\ell$, consider the column

$$C_2(\tau) = \left|\begin{matrix} p(1)^{p(\tau)} \\ \vdots \\ p(\ell)^{p(\tau)} \end{matrix}\right|$$

where $0^0 = 1$ whenever it occurs.

We obtain then

$$(66) \qquad C_2(\tau) = C_1(\tau) + \sum_\lambda \binom{p(\tau)}{p(\lambda)} (-a)^{p(\tau)-p(\lambda)} C_1(p(\lambda)), \quad \forall \ 1 \le \tau \le \ell ,$$

where the sum $\sum_\lambda$ is taken for all $1 \le \lambda \le \ell$ such that $|p(\lambda)| < |p(\tau)|$.

Introducting the determinant

$$\Delta_2 = \det\left(p(\sigma)^{p(\tau)}\right), \text{ where } 1 \le \sigma,\tau \le \ell ,$$

it follows from (66) that $\Delta_2 = \Delta_1$, since $C_2(\tau)$ is the τ-th column in Δ_2 .

We shall now simplify Δ_2 with the help of the function $F : N \times N \to N$ defined by

$$F(h,k) = \begin{cases} 1 & \text{if } k = 0 \\ h(h-1)\dots(h-k+1) & \text{if } k \ge 1 \end{cases}$$

which obviously satisfies the conditions

(67) $F(h,k) = 0 \Leftrightarrow h - k + 1 \leq 0 \Leftrightarrow h < k$

(68) $F(h,h) = h\,!$

Now, for $1 \leq \tau \leq \ell$, we define the column

$$C_3(\tau) = \begin{vmatrix} F(p_1(1)\ ,\ p_1(\tau)) \times \ldots \times F(p_n(1)\ ,\ p_n(\tau)) \\ \vdots \qquad\qquad\qquad\qquad \vdots \\ F(p_1(\ell)\ ,\ p_1(\tau)) \times \ldots \times F(p_n(\ell)\ ,\ p_n(\tau)) \end{vmatrix}$$

Then, it follows that

$$(69) \qquad C_3(\tau) = C_2(\tau) + \Sigma \left(\prod_{j\in J}(-j)\right) \begin{vmatrix} p(1)^{q(\tau)} \\ \vdots \\ p(\ell)^{q(\tau)} \end{vmatrix} , \quad \forall\ 1 \leq \tau \leq \ell ,$$

where

(69.1) the sum Σ is taken for all $J = J_1 \cup \ldots \cup J_n \neq \emptyset$ with $J_i \subset \{1,2,\ldots,p_i(\tau)-1\}$, for $1 \leq i \leq n$,

(69.2) $q(\tau) = p(\tau) - (|J_1|,\ldots,|J_n|)$, with $|J_i|$ denoting the number of elements in J_i .

The relation (69) can obviously be written under the form

$$(70) \qquad C_3(\tau) = C_2(\tau) + \Sigma \left(\prod_{j\in J}(-j)\right) C_2\ (\lambda(\tau,J_1\ ,\ldots,\ J_n))\ , \quad \forall\ 1 \leq \tau \leq \ell ,$$

where (69.1) and (69.2) are still valid and $\lambda(\tau,J_1\ ,\ldots,\ J_n) \in N$ is uniquely defined by

$$p(\lambda(\tau,J_1\ ,\ldots,\ J_n)) = p(\tau) - (|J_1|\ ,\ldots,\ |J_n|)$$

therefore $1 \leq \lambda(\tau,J_1\ ,\ldots,\ J_n) \leq \ell$ and $|\lambda(\tau,J_1\ ,\ldots,\ J_n)| < |p(\tau)|$.

Denoting by Δ_3 the determinant with the columns $C_3(1),\ldots,C_3(\ell)$, the relation (70) implies $\Delta_3 = \Delta_2$ and thus

(71) $\Delta_3 = \Delta_1$

Now, the relation (67) gives for any $1 \leq \sigma,\tau \leq \ell$ the equivalences

$$\prod_{1\leq i\leq n} F(p_i(\sigma)\ ,\ p_i(\tau)) = 0 \Leftrightarrow (\exists\ 1\leq i\leq n : p_i(\sigma) < p_i(\tau)) \Leftrightarrow p(\tau) \not\leq p(\sigma)$$

Therefore, taking into account (68) and the form of the columns $C_3(\tau)$, with $1\leq\tau\leq\ell$, the relation (71) will end the proof ∇∇∇

§8. STRONGER RELATIONS CONTAINING PRODUCTS WITH DIRAC DISTRIBUTIONS

The stronger version of the relations in 2) and 3) in Theorem 1, §4, obtained in the present section is of the following type: Given a locally finite subset $X \subset R^n$, a family of Dirac distribution derivatives $(D^{q_a}\delta_a \mid a \in X)$, with $q_a \in N^n$, and a family $(\psi_a \mid a \in X)$ of functions $\psi_a \in \mathcal{C}^\infty(R^n)$ whose derivatives up to a sufficiently high order vanish in $0 \in R^n$, one obtains within a subclass of the Dirac algebras constructed in §§2-4, the relations:

(72) $$\sum_{a\in X} \psi_a(x-a) \cdot D^{q_a}\delta_a(x) = 0 , \quad x \in R^n .$$

The mentioned Dirac algebras are constructed through a particularization of the procedure in §§2-4. Namely, the family of Dirac classes T_Σ, with $\Sigma \in Z_\delta$, defined in §3, will be replaced by a smaller family which possesses stronger properties.

First, we restrict the representations of the Dirac δ distributions given by Z_δ in §3.

Denote by Z^δ the set of all $\Sigma = (s_x \mid x \in R^n) \in Z_\delta$ satisfying the condition

(73) $\forall \; X \subset R^n$, X locally finite:

(73.1) $(\text{supp } s_x(\nu) \mid x \in X)$ is locally finite, $\forall \; \nu \in N$,

(73.2) $\forall \; x_0 \in R^n$:
$\exists \; V$ neighbourhood of x_0, $\mu \in N$:
$\forall \; \nu \in N, \; \nu \geq \mu$:
$$V \cap \bigcup_{x\in X\setminus\{x_0\}} \text{supp } s_x(\nu) = \emptyset$$

The analog of Corollaries 1 and 2 in 7 is obtained in:

Proposition 5

$Z^\delta \neq \emptyset$ and there exist $\Sigma = (s_x \mid x \in R^n) \in Z^\delta$ such that $s_x \in W_+$, $\forall \; x \in R^n$

Proof

It results from the proof of Corollary 2, §7 ∇∇∇

Now, for $\Sigma = (s_x \mid x \in R^n) \in Z^\delta$, denote by T^Σ the vector subspace in S_0 generated by all the sums

$$\sum_{x\in X} \sum_{\substack{q\in N^n \\ q\leq p_x}} \lambda_{xq} D^q s_x$$

where $X \subset R^n$, X locally finite, $p_x \in N^n$ and $\lambda_{xq} \in C^1$. One can notice that due to (73.1), the definition of T^{Σ} is correct.

And now, the analog of Proposition 2 in §3:

Proposition 6

For each $\Sigma \in Z^{\delta}$, T^{Σ} is a Dirac class which is compatible with the Dirac ideal I^{δ} .

Proof

First, we prove that T^{Σ} is a Dirac class, that is, it satisfies (17.1), (17.2) and (24) in chap. 2. Indeed, the conditions (17.1) and (24) result easily. Assume now $t \in T^{\Sigma}$, $t \notin 0$, then

(74) $$t = \sum_{x \in X} \sum_{\substack{q \in N^n \\ q \le p_x}} \lambda_{xq} D^q s_x$$

where $X \subset R^n$, X locally finite, nonvoid, $p_x \in N^n$ and $\lambda_{xq} \in C^1$. Moreover

(75) $$\exists \; x_o \in X \, , \; q_o \in N^n \, , \; q_o \le p_{x_o} : \lambda_{x_o q_o} \neq 0$$

We show that (17.2) is satisfied for x_o given in (75). Assume, it is false and

(76) $$\exists \; \mu \in N : \forall \; \nu \in N \, , \; \nu \ge \mu : t(\nu)(x_o) = 0$$

then (74) gives

(77) $$\sum_{x \in X} \sum_{\substack{q \in N^n \\ q \le p_x}} \lambda_{xq} D^q s_x(\nu)(x_o) = 0 \, , \quad \forall \; \nu \in N \, , \; \nu \ge \mu$$

Now, the condition (73.2) implies that one can take μ such that (77) will result in

(78) $$\sum_{\substack{q \in N^n \\ q \le p_{x_o}}} \lambda_{x_o q} D^q s_{x_o}(\nu)(x_o) = 0 \, , \quad \forall \; \nu \in N \, , \; \nu \ge \mu$$

Using the same argument as in the proof of Proposition 2, §3, the relation (78) will contradict (75) ending the proof of the fact that T^{Σ} is a Dirac class. It remains to show that T^{Σ} and I^{δ} are compatible. Since, obviously $T_{\Sigma} \subset T^{\Sigma}$ and I^{δ} , T_{Σ} are compatible according to Proposition 2, §3, it suffices to prove that

(79) $$I^{\delta} \cap T^{\Sigma} = 0 \, .$$

Assume therefore $t \in I^{\delta} \cap T^{\Sigma}$, $t \notin 0$, then (74) and (75) hold again, since $t \in T^{\Sigma}$. But $t \in I^{\delta}$ and (2) will again give (76), thus the reasoning above, contradicting (75), will end the proof of (79) ∇∇∇

Based on the above result we proceed to construct the Dirac algebras in which relations of type (72) are valid.

Suppose given $\Sigma \in Z^{\delta}$.

For any ideal I in W , $I \supset I^{\delta}$, compatible vector subspace T in S_o , $T \supset T^{\Sigma}$, vector subspace V in $I \cap V_o$ and vector subspace S_1 in S_o satisfying (16) and (17), one obtains a regularization $(V, T \oplus S_1)$. Then, for any admissible property Q, one can define Dirac algebras according to (18).

The analog of Theorem 1 in §4, stating the validity of (72) within the algebras defined above is obtained in Theorem 9.

First, we shall specify within the algebras the meaning of expressions as in (72), or more general, of the form

$$(80) \qquad \sum_{a \in X} \psi_a(x-a) \cdot \Big(\sum_{\substack{q \in N^n \\ q \leq p_a}} \lambda_{aq} D^q \delta_a \Big)$$

where $X \subset R^n$, X locally finite, $\psi_a \in C^{\infty}(R^n)$, $p_a \in N^n$ and $\lambda_{aq} \in C^1$.

Suppose $H = (h_a \mid a \in X)$ is a family of functions $h_a \in D(R^n)$ such that

(81) $\quad \forall \; a \in X :$

$\exists \; V_a \subset R^n$, V_a neighbourhood of a :

$h_a = 1$ on V_a

(82) $\quad \forall \; a,b \in X$, $a \neq b$:

$\text{supp } h_a \cap \text{supp } h_b = \emptyset$

The existence of such families H results from the fact that X is locally finite.

Obviously, one can define

$$\psi(x) = \sum_{a \in X} h_a(x) \cdot \psi_a(x-a) \; , \quad x \in R^1$$

and then $\Psi \in C^{\infty}(R^n)$. Define $T \in D'(R^n)$ by

$$(83) \qquad T = \sum_{a \in X} \sum_{\substack{q \in N^n \\ q \leq p_a}} \lambda_{aq} D^q \delta_a$$

<u>Lemma 5</u>

Within the algebras A_p , $p \in \bar{N}^n$, the product $\Psi \cdot T$ does not depend on H , provided that (19) is valid.

<u>Proof</u>

Assume $H' = (h'_a \mid a \in X)$ is an other family satisfying (81) and (82) and define $\psi' \in C^{\infty}(R^n)$ by

$$\psi'(x) = \sum_{x \in X} h'_a(x) \cdot \psi_a(x-a) \; , \quad x \in R^1 .$$

We shall prove that

$$(84) \qquad \psi' \cdot T = \psi \cdot T$$

holds within the algebras A_p , $p \in \bar{N}^n$. Indeed, assume $\Sigma = (s_x \mid x \in R^n)$, then the relation (83) above, the inclusion $\mathcal{T}^\Sigma \subset \mathcal{T}$ and 3) in Theorem 2, chap. 1, §8, gives

$$T = \sum_{a \in X} \sum_{\substack{q \in N^n \\ q \leq p_a}} \lambda_{aq} D^q s_a + I^Q(V(p), \mathcal{T} \oplus S_1) \in A_p$$

since $t = \sum_{a \in X} \sum_{\substack{q \in N^n \\ q \leq p_a}} \lambda_{aq} D^q s_a \in \mathcal{T}^\Sigma$ and $T = \langle t, \cdot \rangle$.

Therefore,

$$\psi \cdot T = u(\psi) \cdot t + I^Q(V(p), \mathcal{T} \oplus S_1) \in A_p$$

$$\psi' \cdot T = u(\psi') \cdot t + I^Q(V(p), \mathcal{T} \oplus S_1) \in A_p$$

hence

$$(85) \qquad \psi' \cdot T - \psi \cdot T = u(\psi' - \psi) \cdot t + I^Q(V(p), \mathcal{T} \oplus S_1) \in A_p$$

But, due to (73.2), t satisfies the condition

$$(86) \qquad \begin{array}{l} \forall \ x_0 \in R^n \setminus X : \\ \exists \ V \text{ neighbourhood of } x_0 , \ \mu \in N : \\ \forall \ \nu \in N , \ \nu \geq \mu : \\ \quad V \cap \operatorname{supp} t(\nu) = \emptyset \end{array}$$

while, due to (81), $u(\psi' - \psi) \cdot t$ satisfies the condition

$$(87) \qquad \begin{array}{l} \forall \ x_0 \in X : \\ \exists \ U \text{ neighbourhood of } x_0 : \\ \quad (\psi' - \psi) \cdot t(\nu) = 0 \text{ on } U , \quad \forall \ \nu \in N \end{array}$$

Now, the relations (86) and (87) will imply

$$D^r(u(\psi' - \psi) \cdot t) \in \mathcal{I}^\delta \cap V_0 , \quad \forall \ r \in N^n$$

therefore, (19) and (85) will give (84) ∇∇∇

Lemma 6

Within the algebras A_p , $p \in \bar{N}^n$, the relations hold

$$\left(\sum_{a \in Y} h_a(x) \cdot \psi_a(x-a)\right) \cdot \left(\sum_{a \in Y} \sum_{\substack{q \in N^n \\ q \leq p_a}} \lambda_{aq} D^q \delta_a\right) = \sum_{a \in Y} \psi_a(x-a) \left(\sum_{\substack{q \in N^n \\ q \leq p_a}} \lambda_{aq} D^q \delta_a\right)$$

for any $Y \subset X$, Y finite, provided that (19) is valid.

Proof

Assume $a, b \in Y$, $a \neq b$ and $q \in N^n$, then

(88) $$h_a(x) \cdot \psi_a(x-a) \cdot D^q\delta_b = 0 \in A_p$$

Indeed, assume $\Sigma = (s_x \mid x \in R^n)$, then, according to 3) in Theorem 2, chap. 1, §8, and the inclusion $\mathcal{T}^\Sigma \subset \mathcal{T}$, one obtains

(89) $$h_a(x) \cdot \psi_a(x-a) \cdot D^q\delta_b = u(h_a) \cdot u(\psi_a(\cdot-a)) D^q s_b + I^Q(\mathcal{V}(p),\mathcal{T} \oplus \mathcal{S}_1) \in A_p$$

But, (81) and (82) imply that $b \notin \text{supp } h_a$, therefore, one obtains

(90) $$D^r(u(h_a) \cdot u(\psi_a(\cdot-a)) D^q s_b) \in \mathcal{I}^\delta \cap \mathcal{V}_o , \quad \forall \ r \in N^n$$

taking into account that $s_b \in \mathcal{Z}_b$. Now, the relations (19) and (90) together with (89) will imply (88) ∇∇∇

The Lemmas 5 and 6 suggest within the algebras A_p , $p \in \bar{N}^n$, the following definition of the expressions in (80)

(91) $$\sum_{a\in X} \psi_a(x-a) \cdot \Big(\sum_{\substack{q\in N^n \\ q\le p_a}} \lambda_{aq} D^q\delta_a \Big) = \Big(\sum_{a\in X} h_a(x)\cdot\psi_a(x-a) \Big) \cdot \Big(\sum_{a\in X} \sum_{\substack{q\in N^n \\ q\le p_a}} \lambda_{aq} D^q\delta_a \Big)$$

where $H = (h_a \mid a \in X)$ is any family of functions $h_a \in \mathcal{D}(R^n)$ satisfying (81) and (82).

<u>Theorem 9</u>

In case (19) is valid, the following relations hold in the algebras A_p , with $p \in \bar{N}^n$:

1) $$\sum_{a\in X} \psi_a(x-a) \cdot D^{q_a}\delta_a(x) = 0 \in A_p ,$$

for each $X \subset R^n$, X locally finite, $q_a \in N^n$ and $\psi_a \in \mathcal{C}^\infty(R^n)$ satisfying the condition

(92) $$D^r\psi_a(0) = 0 , \quad \forall \ a \in X , \ r \in N^n , \ r \le p \ \text{ or } \ r \le q_a .$$

In particular, if $p \in N^n$ then

2) $$\sum_{a\in X} (x-a)^{r_a} \cdot D^{q_a}\delta_a(x) = 0 \in A_p ,$$

for each $X \subset R^n$, X locally finite, $r_a , q_a \in N^n$, $r_a \nleq p$ and $r_a \nleq q_a$.

<u>Proof</u>

1) Assume $\Sigma = (s_x \mid x \in R^n)$ and $H = (h_a \mid a \in X)$ is a family of functions $h_a \in \mathcal{D}(R^n)$ satisfying (81) and (82). Then, according to (91)

(93) $$\sum_{a\in X} \psi_a(x-a) \cdot D^{q_a}\delta_a(x) = \Big(\sum_{a\in X} h_a(x)\cdot\psi_a(x-a) \Big) \cdot \Big(\sum_{a\in X} D^{q_a}\delta_a(x) \Big)$$

We shall prove that the right side of the above relation, denoted by S , vanishes in A_p . Indeed, $S \in \mathcal{D}'(R^n)$, thus, taking into account 3) in Theorem 2, chap. 1, §8, and denoting

$$v = u \left(\sum_{a \in X} h_a \cdot \psi_a(\cdot - a) \right) \cdot \sum_{a \in X} D^{q_a} s_a$$

one obtains

$$S = v + I^Q(\mathcal{V}(p), \mathcal{T} \oplus \mathcal{S}_1) \in A_p \,, \tag{94}$$

since $\sum_{a \in X} D^{q_a} s_a \in \mathcal{T}^{\Sigma} \subset \mathcal{T}$.

But $v \in \mathcal{V}_o$ due to the fact that (92) holds for $r \in N^n$, $r \leq q_a$. Therefore

$$D^r v \in \mathcal{V}_o \,, \quad \forall \; r \in N^n \,. \tag{95}$$

Assume now $r \in N^n$, $r \leq p$, then (90) and (5) imply that $D^r v \in \mathcal{I}^{\delta}$ which together with (95) and (19) will give

$$D^r v \in \mathcal{I}^{\delta} \cap \mathcal{V}_o \subset \mathcal{V} \,, \quad \forall \; r \in N^n \,, \quad r \leq p \,.$$

That relation implies $v \in \mathcal{V}(p)$, hence due to (94) the expression in (93) vanishes in A_p .

2) It follows from 1) taking $\psi_a(x) = x^{r_a}$, $\forall \; a \in X$, $x \in R^n$ ∇∇∇

Chapter 6

LINEAR INDEPENDENT FAMILIES OF DIRAC DISTRIBUTIONS

§1. INTRODUCTION

The representations of the Dirac δ distribution used in chapters 4 and 5, were given by weakly convergent sequences of smooth functions satisfying a condition of strong local presence (see (6) in chap. 5, §3). A first consequence of that condition was the nonsymmetry of these representations, implying that the Dirac distribution derivatives $D^q\delta$ of any order $q \in N^n$, are not invariant within the algebras under the transformation of coordinates

$$R^n \ni x \to a \cdot x \in R^n , \quad a = -1 ,$$

(see pct. 2 in Remark 2, chap. 5, §5).

In the present chapter the following stronger result is proved within the algebras containing the distributions in $\mathcal{D}'(R^n)$: Applying to any given derivative $D^q\delta$, $q \in N^n$, of the Dirac distribution the transformations of coordinates

$$R^n \ni x \to a \cdot x \in R^n , \quad a \in R^1 \setminus \{0\} ,$$

one obtaines for pair wise different $a_0 ,\ldots, a_m \in R^1\setminus\{0\}$, linear independent $D^q\delta(a_0x),\ldots,D^q\delta(a_mx)$.

In §4, that result is extended to include also generalized Dirac elements of the form

$$\lim_{a\to\infty} a^n\delta(ax)$$

The above problems are approached within a general framework established in §2, where arbitrary coordinate transformations within the algebras are studied.

§2. COMPATIBLE ALGEBRAS AND TRANSFORMATIONS

Suppose given $\Sigma \in Z_\delta$ (see chap. 5, §§2-4).

Given an ideal I in W , $I \supset I^\delta$, a compatible vector subspace T in S_0 , $T \supset T_\Sigma$, a vector subspace V in $I \cap V_0$ and a vector subspace S_1 in S_0 such that

(1) $$V_0 \oplus T \oplus S_1 = S_0$$

(2) $$U \subset V(p) \oplus T \oplus S_1 , \quad \forall\ p \in \bar{N}^n ,$$

it follows that $(V,T\oplus S_1)$ is a regularization, therefore, one can define for any ad missible property Q the Dirac algebras

(3) $\quad A^Q(V, T \oplus S_1, p)\ , \quad p \in \bar{N}^n$

denoted for the sake of simplicity by A_p .

It will be assumed throughout §§2 and 3 that

(4) $\quad V = I^\delta \cap V_0$

Given a mapping $\alpha : R^n \to R^n$, $\alpha \in C^\infty$ called <u>transformation</u>, we define $\alpha : W \to W$ by

$$(\alpha s)(\nu)(x) = s(\nu)(\alpha(x))\ , \quad \forall\ s \in W\ , \quad \nu \in N\ , \quad \mathbf{x} \in R^1\ ,$$

obtaining thus a homomorphism of the algebra W .

An algebra A_p and the transformation α are called <u>compatible</u>, only if $I^Q(V(p), T \oplus S_1)$ and $A^Q(V(p), T \oplus S_1)$ are invariant of $\alpha : W \to W$.

In that case, one can define the algebra homomorphism $\alpha : A_p \to A_p$ given by

$$\alpha(s + I^Q(V(p), T \oplus S_1)) = \alpha(s) + I^Q(V(p), T \oplus S_1)\ , \quad \forall\ s \in A^Q(V(p), T \oplus S_1)$$

A transformation $\alpha : R^n \to R^n$, $\alpha \in C^\infty$ is called <u>invertible</u>, only if

$$\alpha^{-1} : R^n \to R^n \text{ exists and } \alpha^{-1} \in C^\infty$$

<u>Proposition 1</u>

An algebra A_p and an invertible transformation α are compatible, only if $A^Q(V(p), T \oplus S_1)$ is an invariant of $\alpha : W \to W$.

<u>Proof</u>

The necessity is obvious. Now, the sufficiency. One needs only to show that $I^Q(V(p), T \oplus S_1)$ is an invariant of $\alpha : W \to W$. But, $I^Q(V(p), T \oplus S_1)$ is the ideal in $A^Q(V(p), T \oplus S_1)$ generated by $V(p)$, therefore, due to Lemma 1 below, it is an invariant of $\alpha : W \to W$ $\nabla\nabla\nabla$

<u>Lemma 1</u>

If α is an invertible transformation, then $V(p)$, with $p \in \bar{N}^n$, are invariant of $\alpha : W \to W$.

<u>Proof</u>

Since α is invertible, one obtains easily

(5) $\quad \alpha(V_0) \subset V_0$

The relation

(6) $\alpha(I^{\delta}) \subset I^{\delta}$

is also valid. Indeed, assume $w \in I^{\delta}$ and $x_0 \in R^n$ given. We have to show that αw satisfies in x_0 the conditions (2) and (3) in chap. 5, §2. Denote $x_1 = \alpha(x_0)$. Then w and x_1 satisfy (2) in chap. 5, §2, thus $(\alpha w)(\nu)(x_0) = w(\nu)(x_1) = 0$, for $\nu \in N$ big enough.

But w and x_1 also satisfy (3) in chap. 5, §2, for a certain neighbourhood V_1 of x_1 . Assume V is a neighbourhood of x_0 such that $\alpha(V) \subset V_1$. Now, if $V_1 \cap \text{supp } w(\nu) = \emptyset$ for $\nu \in N$ big enough, then also $V \cap \text{supp } (\alpha w)(\nu) = \emptyset$ for $\nu \in N$ big enough. On the other side, assume that $V_1 \cap \text{supp } w(\nu)$ shrinks to $\{x_1\}$, when $\nu \to \infty$. Now, due to the continuity of α^{-1} it will follow that $V \cap \text{supp}(\alpha w)(\nu)$ shrinks to $\{x_0\}$, when $\nu \to \infty$, and the proof of (6) is completed. The relations (5) and (6) will obviously imply $\alpha(V) \subset V$, since (4) was assumed valid. Then, it is easy to see that $\alpha(V(p)) \subset V(p)$, $\forall\, p \in \bar{N}^n$ ∇∇∇

The result in Proposition 1 above, justifies the following definitions.
Suppose M is a set of invertible transformations. We shall say that a subalgebra A in W has the property P_M , only if A is an invariant of each $\alpha : W \to W$, with $\alpha \subset M$.
Obviously, P_M is an admissible property (see chap. 1, §6).

The algebras (3) will be called <u>M-transform algebras</u>, only if Q is stronger than P_M .

<u>Corollary 1</u>

An M-transform algebra A_p and a transformation $\alpha \in M$ are compatible.

<u>Proof</u>

The subalgebra $A^Q(V(p), T + S_1)$ is invariant of $\alpha : W \to W$ since A_p is an M-transform algebra and $\alpha \in M$ ∇∇∇

<u>§3. LINEAR INDEPENDENT FAMILIES OF DIRAC DISTRIBUTIONS</u>

Denote by M_0 the set of invertible transformations $\alpha_a : R^n \to R^n$, with $a \in R^1 \setminus \{0\}$, defined by $\alpha_a(x) = ax$, $\forall\, x \in R^n$.

<u>Theorem 1</u>

The Dirac distribution derivative transforms $D^q\delta(a_0 x), \ldots, D^q\delta(a_m x)$ with given $q \in N^n$, are linear independent within the M_0-transform algebras A_p , $p \in \bar{N}^n$, provided that $m \leq |p|$ and $a_o, \ldots, a_m \in R^1 \setminus \{0\}$ are pair wise different.

Proof

Assume, it is false and $k_o, \ldots, k_m \subset C^1$ are such that

(7) $$k_o D^q \delta(a_o x) + \ldots + k_m D^q \delta(a_m x) = 0 \in A_p$$

(8) $$\exists \ 0 \leq i \leq m : k_i \neq 0$$

Assume $\Sigma = (s_x \mid x \in R^n)$ then 3) in Theorem 2, chap. 1, §8, and the inclusion $T_\Sigma \subset T$ give

$$D^q \delta = D^q s_o + I^Q(V(p), T \oplus S_1) \in A_p ,$$

therefore

(9) $$k_i D^q \delta(a_i x) = k_i \alpha_{a_i} D^q s_o + I^Q(V(p), T \oplus S_1) \in A_p , \quad \forall \ 0 \leq i \leq m ,$$

since A_p is an M_o-transform algebra. Denote

(10) $$v = \sum_{o \leq i \leq m} k_i \alpha_{a_i} D^q s_o$$

then (7) and (9) imply $v \in I^Q(V(p), T \oplus S_1)$ hence

(11) $$v = \sum_{o \leq j \leq h} v_j \cdot w_j$$

with $v_j \in V(p)$ and $w_j \in A^Q(V(p), T \oplus S_1)$.

Taking into account (4) above, as well as condition (2) in the definition of I^δ in chap. 5, §2, one obtains from (11) the relation

(12) $$\begin{aligned} &\forall \ r \in N^n, \ r \leq p : \\ &\exists \ \mu \in N : \\ &\forall \ \nu \in N, \ \nu \geq \mu : \\ &\quad D^r v(\nu)(0) = 0 \end{aligned}$$

Since $m \leq |p|$, the relations (10) and (12) will give

(13) $$\left(\sum_{o \leq i \leq m} k_i (a_i)^{|r|} \right) D^{q+r} s_o(\nu)(0) = 0 , \quad \forall \ r \in N^n , \ r \leq p , \ |r| \leq m , \ \nu \in N , \ \nu \geq \mu' ,$$

for a suitable $\mu' \in N$.

But $s_o \in Z_o$, therefore

(14) $$\begin{aligned} &\forall \ r \in N^n, \ \sigma \in N : \\ &\exists \ \nu \in N, \ \nu \geq \sigma : \\ &\quad D^r s_o(\nu)(0) \neq 0 \end{aligned}$$

since the matrix $W(s_o)(0)$ is column wise nonsingular (see chap. 5, § 3).

Now, the relations (13) and (14) result in

$$\sum_{0\le i\le m} k_i(a_i)^\ell = 0 \ , \quad \forall \ \ell \in N \ , \quad \ell \le |p|$$

Since $m \le |p|$ and $a_o, \dots, a_m$ are pair wise different, the known property of the Vandermonde determinant will imply $k_o = \dots = k_m = 0$, contradicting (8) ∇∇∇

Corollary 2

The family $(D^q\delta(ax) \mid a \in R^1\setminus\{0\})$ of Dirac distribution derivative transforms with given $q \in N^n$, is linear independent within the M_o-transform algebras A_p , $p \in \bar{N}^n$, $|p| = \infty$.

§4. GENERALIZED DIRAC ELEMENTS

Within $\mathcal{D}'(R^n)$, the operation

$$(15) \qquad \lim_{a\to\infty} a^n \alpha_a S$$

has a meaning for certain distributions S . For instance, if $S = f \in L^1(R^n)$ and $K = \int_{R^n} f(x)dx$, then (15) gives $K\delta$. In case $S = \delta$, one obtains $a^n\alpha_a S = S$, thus (15) will give $S = \delta$.

Within the M_o-transform algebras A_p , $p \in \bar{N}^n$, $|p| = \infty$, the problem of the limit

$$(16) \qquad \lim_{a\to\infty} a^n \alpha_a = \lim_{a\to\infty} a^n\delta(ax) \ , \quad x \in R^n \ ,$$

becomes nontrivial due to Corollary 2 in §3.

A class of algebras, similar to the ones used in §§2 and 3 will be constructed in this section and it will be shown in Theorem 2 that within those algebras, the limit in (16) exists and it is different of $a^n\delta(ax)$, with $a \in R^1\setminus\{0\}$.

The mentioned algebras are constructed by replacing the ideal I^δ defined in chap. 5 §2, with the smaller one I_δ , consisting of all the sequences of smooth functions $w \in I^\delta$ which satisfy the additional condition

(17) $w(\nu)$ vanishes outside of a bounded subset of R^n , provided that $\nu \in N$ is big enough.

It is easy to see that I_δ is indeed an ideal in W , actually a Dirac ideal.

Proposition 2

For each $\Sigma \in Z_\delta$, the Dirac class T_Σ and the Dirac ideal I_δ are compatible.

Proof

It follows from the inclusion $I_\delta \subset I^\delta$ and Proposition 2 in chap. 5, §3 ∇∇∇

Suppose now given $\Sigma \in Z_\delta$.
Given an ideal I in W , $I \supset I_\delta$, a compatible vector subspace T in S_0 , $T \supset T_\Sigma$, a vector subspace V in $I \cap V_0$ and a vector subspace S_1 in S_0 such that

(18) $$V_0 \oplus T \oplus S_1 = S_0$$

(19) $$U \subset V(p) \oplus T \oplus S_1 \ , \quad \forall \ p \in \bar{N}^n \ ,$$

it follows that $(V, T \oplus S_1)$ is a regularization, thus, one can define for any admissible property Q the Dirac algebras

(20) $$A^Q(V, T \oplus S_1, p) \ , \quad p \in \bar{N}^n \ ,$$

denoted for simplicity by A_p .

We shall assume in the sequel that

(21) $$V = I_\delta \cap V_0 \ .$$

Within the above algebras A_p , $p \in \bar{N}^n$, the limit in (16) will be obtained as given by a relation

(22) $$\lim_{a \to \infty} a^n \delta(ax) = t + I^Q(V(p), T \oplus S_1) \in A_p \ ,$$

with $t \in W$, $t(\nu) = (b_\nu)^n \, s(\nu)(b_\nu x)$, $\forall \ \nu \in N$, $x \in R^n$, where $s \in Z_0$ and $b_\nu > 0$, $\lim_{\nu \to \infty} b_\nu = \infty$.

The entities of type (22) will be called <u>generalized Dirac elements</u>.

Using a proof similar to the ones in Proposition 1 and Lemma 1 in §2, one obtains:

<u>Proposition 3</u>

An algebra A_p and an invertible transformation α are compatible, only if $A^Q(V(p), T \oplus S_1)$ is an invariant of $\alpha : W \to W$.

Suppose given $b = (b_\nu \mid \nu \in N)$, with $b_\nu \in R^1 \setminus \{0\}$. Then, one can define the algebra homomorphism $\alpha_b : W \to W$, where

$$(\alpha_b w)(\nu)(x) = |b_\nu|^n \, w(\nu)(b_\nu x) \ , \quad \forall \ w \in W \ , \quad \nu \in N \ , \quad x \in R^n \ .$$

We shall only be interested in the case when

(23) $$\lim_{\nu \to \infty} b_\nu = \pm \infty$$

Since in the definition of compatibility between an algebra A_p and a transformation $\alpha : R^n \to R^n$ given in §2, the transformation appears only through the generated algebra homomorphism $\alpha : W \to W$, it follows that the compatibility has actually been defined between the algebras A_p and algebra homomorphisms of W . In the same way, the definition of M-transform algebras given in §2, will still be correct if M contains besides transformations also algebra homomorphisms of W .

Proposition 4

The algebra A_p and the algebra homomorphism α_b are compatible, only if $A^Q(V(p), T \oplus S_1)$ is an invariant of α_b.

Proof

See the proof of Proposition 1, §2 and the following:

Lemma 2

$V(p)$, with $p \in \bar{N}^n$, are invariant of α_b.

Proof

Assume

$$\lim_{\nu\to\infty} b_\nu = +\infty \tag{24}$$

First we prove

$$\alpha_b(I_\delta) \subset I_\delta \tag{25}$$

Indeed, assume $w \in I_\delta$ and $x_0 \in R^n$ given. First, we notice that due to (24), the sequence of smooth functions $\alpha_b w$ will obviously satisfy (17), since $w \in I_\delta$ satisfies that condition. We shall now show that $\alpha_b w$ satisfies in x_0 the conditions (2) and (3) in chap. 5, §2. Indeed, in case $x_0 \neq 0$, the two conditions result easily from (24), while for $x_0 = 0$ they are obvious.

Now, we prove that

$$\alpha_b(V) \subset V \tag{26}$$

Assume, indeed $v \in V$, then $v \in I_\delta \cap V_0$ therefore, due to (25), it will suffice to show that $\alpha_b v \in V_0$, which is equivalent to proving

$$\lim_{\nu\to\infty} \int_{R^n} v(\nu)(x)\psi(x/b_\nu)dx = 0, \quad \forall \ \psi \in D(R^n) \tag{27}$$

First, we notice that

$$\text{supp } v(\nu) \subset K, \quad \forall \ \nu \in N, \ \nu \geq \mu \tag{28}$$

for suitable $K \subset R^n$, K bounded and $\mu \in N$, since $v \in I_\delta$. Assume now $\chi \in D(R^n)$ such that $\chi = 1$ on K. Due to (28), the relation (27) will hold only if

$$\lim_{\nu\to\infty} \int_{R^n} v(\nu)(x)\chi(x)\psi(x/b \ )dx = 0, \quad \forall \ \psi \in D(R^n) \tag{29}$$

But, for any $\psi \in D(R^n)$, the sequence $(\psi_\nu \mid \nu \in N)$ with $\psi_\nu(x) = \chi(x) \cdot \psi(x/b_\nu)$, $\forall \ x \in R^n$, is convergent in $D(R^n)$ to $\psi(0) \cdot \chi$. Therefore, (29) holds, since $v \in V = I_\delta \cap V_0 \subset V_0$ hence, the sequence $(v(\nu) \mid \nu \in N)$ converges in $D'(R^n)$ to 0 Thus, the relation (26) is proved. It follows then easily that

$$\alpha_b(V(p)) \subset V(p) \ , \qquad \forall \ \ p \in \bar{N}^n \ .$$

The case when $\lim\limits_{\nu\to\infty} b_\nu = -\infty$ is similar $\nabla\nabla\nabla$

Suppose given $b = (b_\nu \mid \nu \in N)$, with $b_\nu \in R^1 \setminus \{0\}$, satisfying (23) and denote $M_b = M_o \cup \{\alpha_b\}$ where M_o was defined in §3.

As noticed above, one can define M_b-transform algebras A_p , with $p \in \bar{N}^n$. According to Propositions 3 and 4, these algebras will be compatible with any $\beta \in M_b$.

And now, the result concerning generalized Dirac elements.

<u>Theorem 2</u>

The Dirac distribution derivative transforms $D^q\delta(a_o x), \ldots, D^q\delta(a_m x)$ and the generalized Dirac element derivative $D^q\alpha_b\delta(x)$, with given $q \in N^n$, are linear independent within the M_b-transform algebras A_p , $p \in \bar{N}^n$, provided that $m \leq |p|$ and $a_o, \ldots, a_m \in R^1 \setminus \{0\}$ are pair wise different.

<u>Proof</u>

Assume, it is false and $k_o, \ldots, k_m$, $k \in C^1$ are such that

(30) $$k_o D^q\delta(a_o x) + \ldots + k_m D^q\delta(a_m x) + k\alpha_b D^q\delta(x) = 0 \in A_p$$

(31) $$k = 0 \Rightarrow \exists \ 0 \leq i \leq m : k_i \neq 0$$

Assume $\Sigma = (s_x \mid x \in R^n)$ then 3) in Theorem 2, chap. 1, §8, and the inclusion $T_\Sigma \subset T$ imply

$$D^q\delta = D^q s_o + I^Q(V(p), T \oplus S_1) \in A_p$$

Since A_p is an M_b-transform algebra, it follows that

(32) $$k_i D^q\delta(a_i x) = k_i \alpha_{a_i} D^q s_o + I^Q(V(p), T \oplus S_1) \in A_p \ , \qquad \forall \ \ 0 \leq i \leq m \ ,$$

(33) $$k D^q\alpha_b\delta(x) = k \cdot t + I^Q(V(p), T \oplus S_1) \in A_p$$

where

(34) $$t(\nu)(x) = |b_\nu|^n D^q s_o(\nu)(b_\nu x) \ , \qquad \forall \ \ \nu \in N \ , \ \ x \in R^n \ .$$

Denote

(35) $$v = \sum_{o \leq i \leq m} k_i \alpha_{a_i} D^q s_o + k \cdot t$$

then (30), (32) and (33) imply $v \in I^Q(V(p), T \oplus S_1)$ thus

(36) $$v = \sum_{o \leq j \leq h} v_j \cdot w_j$$

with $v_j \in V(p)$ and $w_j \in A^Q(V(p), T \oplus S_1)$.

Taking into account (21) above as well as condition (2) in the definition of I^{δ} in chap. 5, §2, the relation (36) will result in

$$(37)\qquad \begin{array}{l} \forall\ r \in N^n,\ r \leq p: \\ \exists\ \mu \in N: \\ \forall\ \nu \in N,\ \nu \geq \mu: \\ \quad D^r v(\nu)(0) = 0 \end{array}$$

But $m \leq |p|$, therefore (35) and (37) will give

$$(38)\qquad \left(\sum_{0\leq i\leq m} k_i (a_i)^{|r|} + k|b_\nu|^{n+|r|} \right) D^{q+r} s_0(\nu)(0) = 0$$

$$\forall\ r \in N^n,\ r \leq p,\ |r| \leq m,\ \nu \in N,\ \nu \geq \mu',$$

for a suitable $\mu' \in N$.

Now, according to (14) in the proof of Theorem 1 in §3, the relation (38) results in

$$(39)\qquad \sum_{0\leq i\leq m} k_i (a_i)^{\ell} + k|b_{\nu_\ell}|^{n+\ell} = 0, \quad \forall\ \ell \in N,\ \ell \leq m,$$

where for any given $\sigma \in N$, one can find suitable $\nu_0, \dots, \nu_\ell \in N$, $\nu_0, \dots, \nu_\ell \geq \sigma$

But $k_0, \dots, k_m, k, a_0, \dots, a_m$ are constant in (39). Therefore, (23) will imply that k must vanish since σ above can be arbitrary. Then, (39) becomes

$$\sum_{0\leq i\leq m} k_i (a_i)^{\ell} = 0, \quad \forall\ \ell \in N,\ \ell \leq m.$$

Since $a_0, \dots, a_m$ are pair wise different, the known property of the Vandermonde determinants will imply $k_0 = \dots = k_m = 0$ which together with $k = 0$ obtained above, will contradict (31) ∇∇∇

Corollary 3

The family $(D^q \delta(ax) \mid a \in R^1 \setminus \{0\})$ of Dirac distribution derivative transforms together with the generalized Dirac element derivative $\alpha_b D^q (x)$ with given $q \in N^n$, are linear independent within the M_b-transform algebras A_p, $p \in \bar{N}^n$, $|p| = \infty$.

Chapter 7

SUPPORT, LOCAL PROPERTIES

§1. INTRODUCTION

An important property of the distribution multiplication, namely its local character, [11], [61], [66-69], [78], which can be formulated as follows

$$\forall \quad S,S',T,T' \in \mathcal{D}'(R^n) \ , \quad E \subset R^n \ , \quad E \neq \emptyset \ , \quad \text{open} :$$

$$\left[\begin{array}{l} *) \quad S \cdot T \ , \quad S' \cdot T' \quad \text{exist in} \quad \mathcal{D}'(R^n) \\ **) \quad S = S' \ , \quad T = T' \quad \text{on} \quad E \end{array} \right] \Rightarrow \quad S \cdot T = S' \cdot T' \quad \text{on} \quad E$$

will be proved valid for the multiplication within the algebras containing the distributions. An extension of the notion of support of a distribution is used in establishing the above, as well as several other local properties of the elements in the algebras.

§2. THE EXTENDED NOTION OF SUPPORT

Suppose P is an admissible property, (V,S') is a P-regularization, Q is an admissible property, such that $Q \leq P$ and $p \in \bar{N}^n$.

Given $S \in A^Q(V,S',p)$ and $E \subset R^n$, we say that S vanishes on E, only if

$$\exists \quad s \in A^Q(V(p),S') :$$

(1)
$$*) \quad S = s + I^Q(V(p),S')$$
$$**) \quad s(\nu) = 0 \quad \text{on} \quad E \ , \ \text{with} \quad \nu \in N \ , \quad \nu \geq \mu \ ,$$

for a certain $\mu \in N$.

The support of S will be the closed subset

(2) $$\text{supp } S = R^n \setminus \{x \in R^n \mid S \quad \text{vanishes on a neighbourhood of} \quad x\} \ .$$

Proposition 1

For functions in $C^\infty(R^n)$ the above notion of support is identical with the usual one.

Proof

It results from (20.2) in §7 and Theorem 2 in §8, chap. 1 ∇∇∇

The local character of the multiplication and addition within the algebras containing the distributions is established in the following two theorems.

Theorem 1

If $S,T \in A^Q(V,S',p)$, then

1) supp $(S + T) \subset$ supp $S \cup$ supp T

2) supp $(S \cdot T) \subset$ supp $S \cap$ supp T

Proof

It follows from a direct verification of the definitions $\nabla\nabla\nabla$

Given $S,S' \in A^Q(V,S',p)$ and $E \subset R^n$, we say that $S = S'$ on E , only if $S - S'$ vanishes on E .

Theorem 2

Suppose $S,S',T,T' \in A^Q(V,S',p)$ and $E \subset R^n$. If $S = S'$ on E and $T = T$ on E , then

1) $S + T = S' + T'$ on E

2) $S \cdot T = S' \cdot T'$ on E

Proof

It follows directly form the definitions $\nabla\nabla\nabla$

An important feature of the above notion of support is pointed out in the case of the algebras where the Dirac δ function is represented by weakly convergent sequences of smooth functions satisfying the condition of strong local presence (see chap. 5, §3 and also chap. 4, §6).

Theorem 3

In the case of the algebras constructed in chapter 5, the derivatives $D^q\delta$ of any order $q \in N^n$ of the Dirac δ distribution, possess the properties:

1) supp $D^q\delta = \{0\}$

2) $D^q\delta$ vanishes on any $E \subset R^n$ such that $0 \notin$ cl E *)

3) $D^q\delta$ does not vanish on $R^n \setminus \{0\}$, provided that the condition (19) in chap. 5, §4 is valid.

*) cl E is the topological closure of $E \subset R^n$

4) $D^q\delta$ does not vanish on $\{0\}$, provided that the condition (27) in chap. 5, §4 is valid.

Proof

Assume $\Sigma = (s_x \mid x \in R^n)$, then 3) in Theorem 2, chap. 1, §8, and the inclusion $T_\Sigma \subset T$ gives for $D^q\delta$ the following representation

(3) $$D^q\delta = D^q s_0 + I^Q(V(p), T \oplus S_1) \in A^Q(V, T \oplus S_1, p), \quad p \in \bar{N}^n$$

Then 1) and 2) result easily from (5) in chap. 5, §3.

3) We notice that in any representation

(4) $$D^q\delta = t + I^Q(V(p), T \oplus S_1) \in A^Q(V, T \oplus S_1, p), \quad p \in \bar{N}^n,$$

t will be a sequence of smooth functions. Therefore, if $D^q\delta$ vanishes on $R^n \setminus \{0\}$, then

(5) $$t(\nu) = 0 \text{ on } R^n, \quad \forall\ \nu \in N, \ \nu \geq \mu,$$

for certain $\mu \in N$. But (5) obviously implies $D^r t \in I^\delta \cap V_0$, $\forall\ r \in N^n$. Thus, due to (19) in chap. 5, §4, one obtains $t \in V(p)$, $\forall\ p \in \bar{N}^n$. Now, (4) will imply $D^q\delta = 0 \in A^Q(V, T \oplus S_1, p)$, $\forall\ p \in \bar{N}^n$, contradicting 1) in Theorem 1, chap. 5, §4.

4) Assume, it is false and there exists a representation (4) with $t \in A^Q(V(p), T \oplus S_1)$ such that

(6) $$t(\nu)(0) = 0, \quad \forall\ \nu \in N, \ \nu \geq \mu',$$

for certain $\mu' \in N$. Denote

(7) $$v = t - D^q s_0$$

then (3) and (4) imply $v \in I^Q(V(p), T \oplus S_1)$ therefore

$$v = \sum_{0 \leq j \leq m} v_j \cdot w_j$$

with $v_j \in V(p)$, $w_j \in A^Q(V(p), T \oplus S_1)$. Now, the condition (27) in chap. 5, §4, will give

$$v_j \in V(p) \subset V \subset I^\delta, \quad \forall\ 0 \leq j \leq m$$

which together with (2) in chap. 5, §2 results in

(8) $$v_j(\nu)(0) = 0, \quad \forall\ 0 \leq j \leq m, \ \nu \in N, \ \nu \geq \mu'',$$

for certain $\mu'' \in N$.

Now, (6), (8) and (7) imply

(9) $$D^q s_0(\nu)(0) = 0, \quad \forall\ \nu \in N, \ \nu \geq \mu$$

with $\mu \in N$ suitably chosen. But, $s_0 \in Z_0$, while (9) obviously contradicts the condition (6) in chap. 5, §3 ∇∇∇

Remark 1

The property in 4), Theorem 3, that the Dirac distribution derivatives $D^q\delta_{x_o}$, with $x_o \in R^n$, $q \in N^n$, do not vanish on $\{x_o\}$ is a consequence of the condition of strong local presence and it is proper for the algebras used in chapters 4, 5 and 6. The 'delta sequences' generally used, [4], [35-41], [53], [68-69], [105-110], [136-137], [162], do not necessarily prevent the vanishing of $D^q\delta_{x_o}$ on $\{x_o\}$.

A characterization of the support of the elements in algebras in terms of the supports of the representing sequences of smooth functions is presented in:

Theorem 4

Suppose $S \in A^Q(V,S',p)$ then

$$\text{supp } S = \cap \text{ cl } \overline{\lim_{\nu\to\infty}} \text{ supp } s(\nu)$$

where the intersection is taken over all the representations

(10) $$S = s + I^Q(V(p),S') \quad A^Q(V,S',p) \quad \text{with} \quad s \in A^Q(V(p),S')$$

Proof

The inclusion $\subset$. Assume s given in (10) and $x \in R^n \setminus \text{cl } \overline{\lim_{\nu\to\infty}} \text{ supp } s(\nu)$. Then

$$V \cap \text{supp } s(\nu) = \emptyset, \quad \forall \quad \nu \in N, \quad \nu \in \mu,$$

for a certain neighbourhood V of x and $\mu \in N$. Now, obviously $x \notin \text{supp } S$.

Conversely, assume $x \in R^n \setminus \text{supp } S$, then there exists an open neighbourhood V of x, such that S vanishes on V. Hence, one can obtain a representation (10), such that $V \cap \overline{\lim_{\nu\to\infty}} \text{ supp } s(\nu) = \emptyset$ ∇∇∇

In the case of the algebras constructed in chap. 6, §4, an additional result on the support can be obtained.

Theorem 5

Suppose given $S \in A^Q(V,T \oplus S_1,p)$ and two representations

$$S = s_1 + I^Q(V(p),T \oplus S_1) = s_2 + I^Q(V(p),T \oplus S_1) \in A^Q(V,T \oplus S_1,p).$$

Then, the subsets in R^n

$$\overline{\lim_{\nu\to\infty}} \text{ supp } s_1(\nu), \quad \overline{\lim_{\nu\to\infty}} \text{ supp } s_2(\nu)$$

differ in at most a finite number of points, provided that $V \subset I_\delta \cap V_o$.

Proof

Obviously $s_1 - s_2 \in I^Q(V(p),T \oplus S_1)$, hence

(8) $$s_1 - s_2 = \sum_{0 \le j \le m} v_j \cdot w_j$$

with $v_j \in V(p)$, $w_j \in A^Q(V(p), T \oplus S_1)$. Now, (27) in chap. 5, §4 implies

$$v_j \in V(p) \subset V \subset I_\delta , \quad \forall \ 0 \le j \le m$$

therefore v_j , with $0 \le j \le m$, satisfy (3) in chap. 5, §2 and (17) in chap. 6, §4. Then, due to (8), $s_1 - s_2$ will also satisfy those two conditions ∇∇∇

Corollary 1

Under the conditions in Theorem 5, if

$$S^m = 0 \in A^Q(V, T \oplus S_1 , p)$$

for certain $m \in N$, $m \ge 1$, then supp S is a finite subset of points in R^n .

Proof

Assume given a representation

$$S = s + I^Q(V(p), T \oplus S_1) , \text{ with } s \in A^Q(V(p), T \oplus S_1)$$

then

$$S^m = s^m + I^Q(V(p), T \oplus S_1) = 0 \in A^Q(V, T \oplus S_1 , p)$$

therefore $s^m \in I^Q(V(p), T \oplus S_1)$.

Using an argument similar to the one in the proof of Theorem 5, it follows that $s^m \in I_\delta$ therefore $\overline{\lim_{\nu \to \infty}} \text{ supp } s^m(\nu)$ is a finite subset of points in R^n . Finally, $\text{supp } s(\nu) = \text{supp } s^m(\nu) , \ \forall \ \nu \in N$ ∇∇∇

Remark 2

In the case of the algebras constructed in chap. 5, the results in Theorem 5 and Corollary 1 will still be valid, provided that finite is replaced by locally finite.

§3. LOCALIZATION

Given $S \in A^Q(V, S', p)$ denote by E_S the set of all open subsets $E \subset R^n$ such that S vanishes on E .

The relation

$$\bigcup_{E \in E_S} E = R^n \setminus \text{supp } S$$

is obvious. In case $S \in D'(R^n)$ and the usual notions of vanishing and support for distributions are used, the corresponding set E_S has the known property that any union of sets in E_S is again a set in E_S.

In particular

$$R^n \setminus \operatorname{supp} S = \bigcup_{E \in E_S} E \in E_S$$

A first problem approached in the present section is the extension of the above property to the algebras containing the distributions. In this respect, several results on the structure of E_S will be given.

Theorem 6

Suppose $S \in A^Q(V,S',p)$ and E_1, $E_2 \in E_S$. If $d(E_1\setminus E_2, E_2\setminus E_1) > 0$ *) then $E_1 \cup E_2 \in E_S$.

Proof

Assume s_1, $s_2 \in A^Q(V(p),S')$ such that

(9) $$S = s_i + I^Q(V(p),S') \in A^Q(V,S',p), \quad \forall\ 1 \le i \le 2,$$

(10) $$s_i(\nu) = 0 \text{ on } E_i, \quad \forall\ 1 \le i \le 2, \ \nu \in N, \ \nu \ge \mu,$$

for certain $\mu \in N$. Denote $v = s_1 - s_2$, then (9) implies

(11) $$v \in I^Q(V(p),S')$$

Further, one obtains for $\nu \in N$, $\nu \ge \mu$

(12) $$v(\nu)(x) = \begin{cases} s_1(\nu)(x) - s_2(\nu)(x) & \text{if } x \in R^n \setminus (E_1 \cup E_2) \\ s_1(\nu)(x) & \text{if } x \in E_2 \setminus E_1 \\ -s_2(\nu)(x) & \text{if } x \in E_1 \setminus E_2 \\ 0 & \text{if } x \in E_1 \cap E_2 \end{cases}$$

According to Lemma 1 below, there exists $\psi \in C^\infty(R^n)$, such that $\psi = -1/2$ on $E_2\setminus E_1$ and $\psi = 1/2$ on E_1/E_2. Denote $w = u(\psi) \cdot v$, then (11) implies $w \in I^Q(V(p),S')$, thus denoting $s = (s_1+s_2)/2 + w$, the relation (9) will give

(13) $$S = s + I^Q(V(p),S') \in A^Q(V,S',p), \quad s \in A^Q(V,S',p)$$

But, due to (12), it follows obviously that

(14) $$s(\nu) = 0 \text{ on } E_1 \cup E_2, \quad \forall\ \nu \in N, \ \nu \ge \mu$$

Now, (13) and (14) will imply $E_1 \cup E_2 \in E_S$ ∇∇∇

---

*) $d(\ ,\ )$ is the Euclidian distance on R^n and $d(E,F) = \inf\{d(x,y) \mid x \in E, y \in F\}$ for $E,F \subset R^n$

Lemma 1

Suppose $F \subset G \subset R^n$ such that $d(F, R^n \backslash G) > 0$, then there exists $\psi \in C^\infty(R^n)$ with the properties

1) $0 \le \psi \le 1$ on R^n

2) $\psi = 1$ on F

3) $\psi = 0$ on $R^n \backslash G$

4) $\psi \in D(R^n)$ if F bounded

Proof

Define $\chi : R^n \times (0,\infty) \to R^1$ by

$$\chi(x,\varepsilon) = \begin{cases} K_\varepsilon \cdot \exp\left(\varepsilon^2/(||x||^2-\varepsilon^2)\right) & \text{if } ||x|| < \varepsilon \\ 0 & \text{if } ||x|| \ge \varepsilon \end{cases}$$

where $K_\varepsilon = 1 / \int_{||x||<\varepsilon} \exp\left(\varepsilon^2/(||x||^2-\varepsilon^2)\right)dx$

Assume $0 < \varepsilon < d(F, R^n \backslash G)/2$ and define $\psi : R^n \to R^1$ by

$$\psi(x) = \int_{F(\varepsilon)} \chi(x-y,\varepsilon)\,dy$$

where $F(\varepsilon) = \{y \in R^n \mid d(y,F) \le \varepsilon\}$

It can easily be seen that ψ is the required function ∇∇∇

Corollary 2

Suppose $S \in A^Q(V,S',p)$ and $E, F \in E_S$. If

1) $\mathrm{cl}\, E \cap \mathrm{cl}\, F = \emptyset$

2) F bounded

then $E \cup F \in E_S$

Proof

The sets E and F satisfy the conditions in Theorem 6 ∇∇∇

Theorem 7

Suppose $S \in A^Q(V,S',p)$ and $E_1, E_2 \in E_S$. If $E_1' \subset E_1$ and $d(E_1', R^n \backslash E_1) > 0$ then $E_1' \cup E_2 \in E_S$.

Proof

We shall use the notations in the proof of Theorem 6.
According to Lemma 1, there exists $\psi \in C^\infty(R^n)$ such that $\psi = -1/2$ on E_2/E_1 and $\psi = 1/2$ on E_1'. Then, it can be seen that the relation (14) becomes

(15) $\quad s(\nu) = 0$ on $E_1' \cup E_2$, $\forall \; \nu \in N$, $\nu \geq \mu$,

therefore, the relation (13) will imply $E_1' \cup E_2 \in E_S$ ∇∇∇

Corollary 3

Suppose $S \in A^Q(V,S',p)$ and $E \subset R^n \setminus \text{supp } S$, E open. If

1) $\text{cl } E \cap \text{supp } S = \emptyset$

2) E bounded

then $E \in E_S$

Proof

Assume $K \subset R^n \setminus \text{supp } S$, K compact, $K \supset E$. It follows from (2) that

$$\forall \; x \in K : \exists \; \varepsilon_x > 0 : B(x,\varepsilon_x) \in E_S \quad ^{*)}$$

Assume $x_0, \ldots, x_m \in K$ pair wise different, such that

(16) $$K \subset \bigcup_{o \leq i \leq m} B(x_i, \varepsilon_{x_i}/2)$$

If $m = 0$ then $E \subset K \subset B(x_0, \varepsilon_{x_0}/2)$ and the proof is completed.

Assume $m = 1$. Denote

$$E_1 = B(x_1, \varepsilon_{x_1}), \quad E_2 = B(x_0, \varepsilon_{x_0}), \quad E_1' = B(x_1, \varepsilon_{x_1}/2)$$

then E_1, E_2 and E_1' fulfil the conditions in Theorem 7, therefore $E_1' \cup E_2 \in E_S$ and due to (16) the proof is again completed.

Assume $m = 2$. Denote

$$E_1 = B(x_2, \varepsilon_{x_2}), \quad E_2 = B(x_0, \varepsilon_{x_0}) \cup B(x_1, \varepsilon_{x_1}/2),$$
$$E_1' = B(x_2, \varepsilon_{x_2}/2)$$

then S vanishes on E_1 and as seen above, also on E_2. Moreover, E_1, E_2 and E_1' fulfil the conditions in Theorem 7, therefore $E_1' \cup E_2 \subset E_S$ and due to (16) the proof is completed again.

The above procedure can be used for any $m \in N$, $m \geq 3$ ∇∇∇

---

*) $B(x,\varepsilon) = \{y \in R^n \mid ||y-x|| < \varepsilon\}$ for $x \in R^n$, $\varepsilon > 0$

Corollary 4

Suppose $S \in A^Q(V,S',p)$ and $H \subset R^n$, then supp $S \subset H$, only if

$$\forall \; K \subset R^n \setminus H \, , \; K \text{ compact:}$$

(17) $$\exists \; E \in E_S : \quad K \subset E$$

Proof

Assume (17) and $x \in R^n \setminus H$ then, $x \in E$ for a certain $E \in E_S$. Therefore $x \notin$ supp S since E is open. The converse results directly from Corollary 3 taking into account that supp S is closed ∇∇∇

Corollary 5

Suppose $S \in A^Q(V,S',p)$, then supp $S = \emptyset$, only if

$$\forall \; E \subset R^n \, , \; E \text{ open, bounded:} \quad E \in E_S$$

Proof

It follows directly from Corollary 3 ∇∇∇

Theorem 8

Suppose $S \in A^Q(V,S',p)$ and $F \subset R^n$, F closed. Then

$$\left(\begin{array}{l} \forall \; \psi \in D(R^n \setminus F) \; : \\ \quad \psi \cdot S = 0 \in A^Q(V,S',p) \end{array} \right) \Rightarrow \text{supp } S \subset F$$

In the case of sectional algebras (see chap. 1, §6) the converse implication is also valid.

Proof

Assume $x \in R^n \setminus F$. Since F is closed, it follows that there exists $\psi \in D(R^n \setminus F)$ and a neighbourhood V of x such that $\psi = 1$ on V . Then, due to the hypothesis $\psi \cdot S = 0 \in A^Q(V,S',p)$, therefore, given any representation

(18) $$S = s + I^Q(V(p),S') \in A^Q(V,S',p) \text{ , with } s \in A^Q(V(p),S')$$

one obtains

(19) $$u(\psi) \cdot s \in I^Q(V(p),S')$$

But, (18) and (19) imply

(20) $$S = u(1-\psi) \cdot s + I^Q(V(p),S') \in A^Q(V,S',p)$$

Denoting $t = u(1-\psi) \cdot s$, it follows that $t(\nu) = 0$ on V , $\forall \nu \in N$. Then (20) will imply $x \notin \operatorname{supp} S$ and the first part of Theorem 8 is proved.

Assume now, in the case of sectional algebras the inclusion $\operatorname{supp} S \subset F$ and $\psi \in \mathcal{D}(R^n \setminus F)$. Since $\operatorname{supp} \psi$ is compact, Corollary 4 implies the existence of $E \in \mathcal{E}_S$ such that

(21) $\operatorname{supp} \psi \subset E$

Due to the fact that S vanishes on E , one can assume that s in (18) satisfies the condition

(22) $s(\nu) = 0$ on E , $\forall\ \nu \in N$, $\nu \geq \mu$,

for certain $\mu \in N$. Now, the presence of sectional algebras makes it possible to assume $\mu = 0$ in (22). Then, (21) and (22) will result in $u(\psi) \cdot s \in \mathcal{O}$ hence (18) will imply $\psi \cdot S = 0 \in A^Q(V, S', p)$ ∇∇∇

Two <u>decomposition</u> rules for the elements of the algebras, corresponding to components of their supports are given now.

<u>Theorem 9</u>

In the case of <u>sectional algebras</u> suppose $S \in A^Q(V, S', p)$. If $\operatorname{supp} S = F \cup K$ with F closed, K compact and $F \cap K = \emptyset$, then the decomposition holds

$$S = S_F + S_K$$

for certain S_F , $S_K \in A^Q(V, S', p)$ satisfying the conditions

1) $\operatorname{supp} S_F \cap \operatorname{supp} S_K = \emptyset$

2) $K \cap \operatorname{supp} S_F = \emptyset$

3) $F \cap \operatorname{supp} S_K = \emptyset$ and $\operatorname{supp} S_K$ compact

<u>Proof</u>

Assume G_1 , G_2 , G_3 , $G_4 \subset R^n$ such that $K \subset G_1$, $\operatorname{cl} G_1 \subset G_2$, $\operatorname{cl} G_2 \subset G_3$, $\operatorname{cl} G_3 \subset G_4$, $\operatorname{cl} G_4 \cap F = \emptyset$ and G_4 is bounded. Denote $K_1 = (\operatorname{cl} G_4) \setminus G_1$, then K_1 is compact and $K_1 \cap \operatorname{supp} S = \emptyset$. According to Corollary 4, there exists $E \in \mathcal{E}_S$ such that $K_1 \subset E$. Then, for a certain representation

$$S = s + I^Q(V(p), S') \in A^Q(V, S', p) \ , \quad \text{with} \quad s \in A^Q(V(p), S') \ ,$$

one obtains

$$s(\nu) = 0 \quad \text{on} \quad E \ , \quad \forall \ \nu \in N \ , \quad \nu \geq \mu$$

with suitably chosen $\mu \in N$. But the case of sectional algebras allows the choice of $\mu = 0$. Now, Lemma 1 grants the existence of $\psi_F \in C^\infty(R^n)$ and $\psi_K \in \mathcal{D}(R^n)$ such that $\psi_F = 1$ on $R^n \setminus G_4$, $\psi_F = 0$ on $\operatorname{cl} G_3$, $\psi_K = 1$ on $\operatorname{cl} G_1$ and $\psi_K = 0$ on $R^n \setminus G_2$.

Then, obviously $u(\psi_F + \psi_K) \cdot s = s$. Defining $S_F = u(\psi_F) \cdot s + I^Q(V(p),S')$ and $S_K = u(\psi_K) \cdot s + I^Q(V(p),S')$, one obtains $S = S_F + S_K$ and the properties 1), 2) and 3) will be satisfied ∇∇∇

In the one dimensional case $n = 1$, a stronger decomposition can be obtained. First, a notion of separation for pairs of subsets in R^1 . Two subsets $F , L \subset R^1$ are called finitely separated, only if there exists a finite number of intervals covering $F \cup L$

$$(-\infty , c_o) , (c_o , c_1) , \dots , (c_m , \infty) , \quad m \in N ,$$

such that no interval contains elements of both F and L while successive intervals do not contain elements of the same set F or L .

Theorem 10

In the case of sectional algebras suppose $S \in A^Q(V,S',p)$. If $\text{supp } S = F \cup L$ with F , L disjoint, closed and finitely separated, then the decomposition holds

$$S = S_F + S_L$$

for certain $S_F , S_L \in A^Q(V,S',p)$ satisfying the conditions

1) $\text{supp } S_F \cap \text{supp } S_L = \emptyset$

2) $L \cap \text{supp } S_F = F \cap \text{supp } S_L = \emptyset$

Proof

Since F , L are closed, there exists $\varepsilon > 0$ such that

$$F \cup L \subset (-\infty,c_o-\varepsilon) \cup (c_o+\varepsilon,c_1-\varepsilon) \cup \dots \cup (c_m+\varepsilon,\infty)$$

Denote

$$K = \bigcup_{0\leq i\leq m} [c_i-\varepsilon,c_i+\varepsilon]$$

then K compact and $K \subset R^1 \setminus \text{supp } S$. According to Corollary 4, there exists $E \in \mathcal{E}_S$ such that $K \subset E$. Assume a representation

$$S = s + I^Q(V(p),S') \in A^Q(V,S',p) , \quad \text{with } s \in A^Q(V(p),S')$$

such that

$$s(\nu) = 0 \text{ on } E , \quad \forall \ \nu \in N , \ \nu \geq \mu$$

for certain $\mu \in N$. But, one can assume $\mu = 0$, due to the presence of sectional algebras. Further, Lemma 1 implies the existence of $\psi_o , \dots , \psi_{m+1} \in C^\infty(R^1)$ such that

$$\psi_o = 1 \text{ on } (-\infty,c_o-\varepsilon/2] , \quad \psi_1 = 1 \text{ on } [c_1+\varepsilon/2,c_2-\varepsilon/2] , \dots$$

$$\dots , \psi_{m+1} = 1 \text{ on } [c_m+\varepsilon/2,\infty)$$

and

$$\psi_0 = 0 \quad \text{on } [c_0-\varepsilon/3, +\infty) \ , \quad \psi_1 = \text{on } (-\infty, c_1+\varepsilon/3] \cup [c_2-\varepsilon/3, \infty) \ , \ \ldots$$
$$\ldots \ , \ \psi_{m+1} = 0 \quad \text{on } (-\infty, c_m+\varepsilon/3]$$

Denote

$$J_0 = (-\infty, c_0] \ , \quad J_1 = [c_0, c_1] \ , \ \ldots \ , \ J_{m+1} = [c_m, \infty)$$

and define

$$\psi_F = \sum_{\substack{o \le i \le m+1 \\ F \cap J_i = \emptyset}} \psi_i \ , \quad \psi_L = \sum_{\substack{1 \le i \le m+1 \\ L \cap J_i = \emptyset}} \psi_i$$

It can easily be seen that $u(\psi_F + \psi_L)s = s$. Defining $S_F = u(\psi_F) \cdot s + I^Q(V(p),S')$ and $S_L = u(\psi_L) \cdot s + I^Q(V(p),S')$, the required properties will follow easily ▽▽▽

§4. THE EQUIVALENCE BETWEEN $S = 0$ AND $\text{supp } S = \emptyset$

If $S = 0 \in A^Q(V,S',p)$ then, obviously $\text{supp } S = \emptyset$. A result on the converse implication is given in:

Corollary 6

In the case of sectional algebras suppose $S \in A^Q(V,S',p)$. If

1) $\text{supp } S = \emptyset$

2) S vanishes outside of a bounded subset of R^n ,

then $S = 0 \in A^Q(V,S',p)$.

Proof

Assume $B \subset R^n$, B bounded, such that S vanishes on $R^n \setminus B$. Then, there exists a representation

$$S = s + I^Q(V(p),S') \in A^Q(V,S',p) \ , \quad \text{with } s \in A^Q(V(p),S') \ ,$$

such that

$$s(\nu) = 0 \quad \text{on } R^n \setminus B \ , \quad \forall \ \nu \in N \ , \quad \nu \ge \mu \ ,$$

for a certain $\mu \in N$. Due to the presence of sectional algebras, one can assume $\mu = 0$ Assume $\psi \in D(R^n)$ such that $\psi = 1$ on B , then obviously $u(\psi) \cdot s = s$, that is

(23) $\qquad \psi \cdot S = S \in A^Q(V,S',p)$

But $\text{supp } \psi \cap \text{supp } S = \emptyset$ and $\text{supp } S$ is closed, therefore Theorem 8 will imply $\psi \cdot S = 0 \in A^Q$. The relation (23) completes the proof ▽▽▽

Chapter 8

NECESSARY STRUCTURE OF THE DISTRIBUTION MULTIPLICATIONS

§1. INTRODUCTION

The algebras containing the distributions in $\mathcal{D}'(R^n)$ were constructed as sequential completions of the smooth functions on R^n and resulted as quotients A/I, where A is a subalgebra in $W = N \to C^\infty(R^n)$, while I is an ideal in A.

That construction can naturally be placed within the framework of the theory of algebras of continuous functions, noticing that W itself is a subalgebra in $C^0(N \times R^n, C^1)$. As known in that context, an essential feature of the ideals I is their connection with certain zero-filters generated by subsets of $N \times R^n$ on which the functions in I vanish.

In the present chapter, specific connections between ideals and zero-filters will be established in the case of the quotient constructions giving the algebras containing the distributions. In that way, an information on the necessary structure of the distribution multiplications will be obtained.

§2. ZERO SETS AND FAMILIES

For a sequence of smooth functions $w \in W$ denote

$$Z(w) = \{ (\nu, x) \in N \times R^n \mid w(\nu)(x) = 0 \}$$

and call it the zero set of w.

For a set of sequences of smooth functions $H \subset W$ denote

$$Z(H) = \{ Z(t) \mid t \in H \}$$

and call it the zero family of H.

A standard argument in algebras of functions gives:

Theorem 1

Suppose (V, S') is a regularization and denote by I the ideal in W generated by V. Then $Z(I)$ and in particular, $Z(V)$ are filter generators on $N \times R^n$.

Proof

First we prove the relation

(1) $\quad Z(v) \neq \emptyset \ , \quad \forall \ v \in \mathcal{I} = I(V,W)$

Assume it is false, then $v(\nu)(x) \neq 0$, $\forall \ \nu \in N$, $x \in R^n$, therefore, one can define $w \in W$ by $w(\nu)(x) = 1/v(\nu)(x)$, $\forall \ \nu \in N$, $x \in R^n$. Thus $u(1) = v \cdot w \in I(V,W)$ which results in $I(V,W) = W$. Then (20.3) and (20.1) in chap. 1, will contradict each other.

Now, we can prove

(2) $\quad Z(v_o) \cap \ldots \cap Z(v_h) \neq \emptyset \ , \quad \forall \ v_o, \ldots, v_h \in I(V,W)$

Indeed, define $w \in W$ by $w = v_o \cdot v_o^* + \ldots + v_h \cdot v_h^*$ where v_i^* is the complex conjugate of v_i . Then, obviously

(3) $\quad Z(w) = Z(v_o) \cap \ldots \cap Z(v_h)$

But, $w \in I(V,W)$, since $v_o, \ldots, v_h \in I(V,W)$. Therefore, (3) and (1) will imply (2) ∇∇∇

The following three theorems reformulate in terms of zero sets the results in Theorems 6, 7 and 8 in chap. 1, §10.

Theorem 2

Suppose given a local type regularization (V,S') , with $V \oplus S'$ sectional invariant. Then

$$Z(v) \cap (\{\mu,\mu+1,\ldots\} \times G) \text{ is infinite}$$

for any $v \in V$, $\mu \in N$ and $G \subset R^n$, $G \neq \emptyset$, open.

Theorem 3

Suppose given a regularization (V,S') such that $V \oplus S'$ is sectional invariant. Then, the zero set $Z(v)$ of each $v \in V$ has the property:

$$Z(v) \cap \bigcup_{\mu \leq \nu < \infty} (\{\nu\} \times \operatorname{supp} t(\nu)) \text{ infinite}, \quad \forall \ t \in V \oplus S' \ , \quad t \notin V \ , \quad \mu \in N$$

Theorem 4

Suppose (V,S') is a regularization. Then, the zero set $Z(v)$ of each $v \in V$ has the property:

$$Z(v) \cap \bigcup_{\nu \in N} (\{\nu\} \times \operatorname{supp} t(\nu)) \neq \emptyset \ , \quad \forall \ t \in V \oplus S' \ , \quad t \notin V$$

§3. ZERO SETS AND FAMILIES AT A POINT

The results in Theorems 1, 2, 3, and 4 in §2, will be strengthened in the present section.

Given $w \in W$, $H \subset W$ and $x \in R^n$, call

$$Z_x(w) = \{ \nu \in N \mid w(\nu)(x) = 0 \}$$

the zero set of w at x and call

$$Z_x(H) = \{ Z_x(t) \mid t \in H \}$$

the zero family of H at x .

It will be shown in Corollaries 1, 2 and 3 that for a large class of P-regularizations (V,S') , the zero families $Z_x(V)$ of V at any $x \in R^n$ are filter generators on N

First, several notations.

Suppose E is a set of subsets in R^n and denote

$$W_E = \left\{ w \in W \;\middle|\; \begin{array}{l} \forall \ E \in E : \\ \exists \ x \in E : \\ \quad Z_x(w) \neq \emptyset \end{array} \right\}$$

Call E hereditary, only if

$$\begin{array}{l} \forall \ x \in E \in E : \\ \quad E \setminus \{x\} \neq \emptyset \Rightarrow E \setminus \{x\} \in E \end{array}$$

Denote by E_f and E_c the set of all nonvoid and finite, respectively countable subsets in R^n . Obviously E_f and E_c are hereditary.

Theorem 5

Suppose given a vector subspace V in W , satisfying for a certain $E \subset E_c$ the condition

$$V \subset W_E$$

Then, for any $v_0, \ldots, v_h \in V$ the property holds

$$\begin{array}{l} \forall \ E \in E : \\ \exists \ x \in E : \\ \quad Z_x(v_0) \cap \ldots \cap Z_x(v_h) \neq \emptyset \end{array}$$

If E is hereditary, then the stronger property results

$$\begin{array}{l} \forall \ E \in E : \\ \exists \ E' \subset E : \\ \quad 1) \ \ \text{car } E' = \text{car } E \\ \quad 2) \ \ Z_x(v_0) \cap \ldots \cap Z_x(v_h) \neq \emptyset , \quad \forall \ x \in E' \end{array}$$

Proof

It follows from Lemma 1 below ∇∇∇

Corollary 1

Suppose the P-regularization (V, S') satisfies the condition

(4) $$Z_x(v) \neq \emptyset \,, \quad \forall \; v \in V \,, \quad x \in R^n$$

Then, the zero family $Z_x(V)$ of V at any $x \in R^n$ is a filter generator on N.

Proof

It is easy to notice that (4) is equivalent with $V \subset W_{E_f}$ ∇∇∇

Corollary 2

If under the conditions in Corollary 1, V is sectional invariant, then the zero family $Z_x(V)$ of V at any $x \in R^n$, generates a filter of infinite subsets of N.

Proof

Assume it is false and $x \in R^n$, $v_0, \dots, v_h \in V$ and $\mu \in N$ such that

(5) $$Z_x(v_0) \cap \dots \cap Z_x(v_h) \subset \{0, \dots, \mu\}$$

Define $v'_0, \dots, v'_h \in W$ by

(6) $$v'_i(\nu)(y) = \begin{cases} 1 & \text{if } \nu \leq \mu \\ v_i(\nu)(y) & \text{if } \nu \geq \mu + 1 \end{cases}$$

Then $v'_i \in V$, since $v_i \in V$ and V is sectional invariant. Therefore

(7) $$Z_x(v'_0) \cap \dots \cap Z_x(v'_h) \neq \emptyset$$

due to Corollary 1.

But, (5) implies that

$$\forall \; \nu \in N \,, \; \nu \geq \mu + 1 :$$
$$\exists \; i_\nu \in \{0, \dots, h\} :$$
$$v_{i_\nu}(\nu)(x) \neq 0$$

which together with (6) is contradicting (7) ∇∇∇

Call a subset M of N sectional, only if

$$\{\mu+1 \,, \mu+2 \,, \dots\} \subset M$$

for a certain $\mu \in N$.

Corollary 3

If under the conditions in Corollary 1, V is subsequence invariant (see chap. 1, §6), then the zero family $Z_x(V)$ of V at any $x \in R^n$, generates a filter of sectional subsets of N.

Proof

It suffices to show that the sets $Z_x(v)$, with $v \in V$ and $x \in R^n$, are all sectional in N. Assume it is false and $v \in V$ and $x \in R^n$ such that $Z_x(v)$ is not sectional in N. Then, there exists an infinite subset M of N, such that

(8) $$Z_x(v) \cap M = \emptyset$$

Assume $M = \{\mu_0, \mu_1, \ldots\}$ and define $v' \in W$ by

(9) $$v'(\nu)(y) = v(\mu_\nu)(y), \quad \forall \ \nu \in N, \ y \in R^n$$

Then $v' \in V$, since $v \in V$ and V is subsequence invariant. Therefore

(10) $$Z_x(v') \neq \emptyset$$

due to Corollary 1.

But, (8) implies that

$$\forall \ \nu \in N : \quad v(\mu_\nu)(x) \neq 0$$

which together with (9) is contradicting (10) $\nabla\nabla\nabla$

And now, a lemma of a general interest, used in the proof of Theorem 5, is presented.

Given a nonvoid set X, denote by W the set of all functions $w : N \times X \to C^1$.

For $w \in W$ and $x \in X$ denote

$$Z_x(w) = \{ \nu \in N \mid w(\nu,x) = 0 \}$$

For a set Y of nonvoid and countable subsets in X, denote

$$W_Y = \left\{ w \in W \;\middle|\; \begin{array}{l} \forall \ Y \in Y : \\ \exists \ y \in Y : \\ \quad Z_y(w) \neq \emptyset \end{array} \right\}$$

Lemma 1

Suppose V is a vector subspace in W and $V \subset W_Y$. Then, for any $v_0, \ldots, v_h \in V$, the relation holds

(11) $$\begin{array}{l} \forall \ Y \in Y : \\ \exists \ y \in Y : \\ \quad Z_y(v_0) \cap \ldots \cap Z_y(v_h) \neq \emptyset \end{array}$$

If $\mathcal{Y}$ is hereditary, then (11) obtains the stronger form

$$(12)\quad \begin{aligned} &\forall \ Y \in \mathcal{Y} : \\ &\exists \ Y' \subset Y : \\ &\quad (12.1) \quad \text{car } Y' = \text{car } Y \\ &\quad (12.2) \quad Z_y(v_o) \cap \ldots \cap Z_y(v_h) \neq \emptyset , \quad \forall \ y \in Y' \end{aligned}$$

Proof

First, we prove (11). Assume it is false and $Y \in \mathcal{Y}$ such that

$$Z_y(v_o) \cap \ldots \cap Z_y(v_h) = \emptyset , \quad \forall \ y \in Y$$

Then

$$(13)\quad \begin{aligned} &\forall \ \nu \in N , \ y \in Y : \\ &\exists \ i_{\nu,y} \in \{0,\ldots,h\} : \\ &\quad v_{i_{\nu,y}}(\nu,y) \neq 0 \end{aligned}$$

Define

$$R^{h+1} \ni \lambda = (\lambda_o ,\ldots, \lambda_h) \rightarrow v_\lambda = \lambda_o v_o + \ldots + \lambda_h v_h \in V$$

and

$$N \times Y \ni (\nu,y) \rightarrow \Lambda_{\nu,y} = \{ \lambda \in R^{h+1} \mid v_\lambda(\nu,y) = 0 \}$$

It is easy to notice that $\Lambda_{\nu,y}$, with $(\nu,y) \in N \times Y$, are vector subspaces in R^{h+1}. Moreover

$$(14)\quad \Lambda_{\nu,y} \subsetneqq R^{h+1} , \quad \forall \ (\nu,y) \in N \times Y$$

Indeed, denoting $\lambda = (0,\ldots,0,1,0,\ldots,0) \in R^{h+1}$, with 1 in the $i_{\nu,y}+1$ -th position, one obtains $\lambda \notin \Lambda_{\nu,y}$ due to (13).

Now, (14) and the Baire category argument will give

$$\bigcup_{\substack{\nu\in N \\ y\in Y}} \Lambda_{\nu,y} \subsetneqq R^{h+1}$$

since Y is countable.

Assume therefore

$$\lambda \in R^{h+1} \setminus \bigcup_{\substack{\nu\in N \\ y\in Y}} \Lambda_{\nu,y}$$

then

$$(15)\quad v_\lambda(\nu,y) \neq 0 , \quad \forall \ \nu \in N , \ y \in Y$$

But $v_\lambda \in V \subset W_Y$ therefore (15) is contradicted and (11) is proved.

Assume $\mathcal{Y}$ is hereditary and take $Y \in \mathcal{Y}$. Then (11) implies the existence of $y \in Y$ such that $Z_y(v_o) \cap \ldots \cap Z_y(v_h) \neq \emptyset$. Denote $Y_1 = Y \setminus \{y\}$. If $Y_1 = \emptyset$ then taking

$Y' = Y$, the proof is completed. If $Y_1 \neq \emptyset$ then $Y_1 \in \mathcal{Y}$ due to the fact that $\mathcal{Y}$ is hereditary. Now, (11) can be applied to Y_1 , etc. ∇∇∇

REFERENCE

1. Acker A., Walter W.: The quenching problem for nonlinear parabolic differential equations. Springer Lecture Notes in Mathematics, vol. 564, 1-12

2. Alexandrov A.D.: Dirichlet's problem for the equation $\det(z_{i_j}) = \Phi(z_1,\dots,z_n)$ Vestinik Leningrad Univ. Ser. Mat. Mek. Astr. 13,1,1958

3. Amrein W.O., Georgescu V.; Strong asymptotik completeness of wave operators for highly singular potentials. Helv. Phys. Acta, 47,1974, 517-533

4. Antosik P., Mikusinski J., Sikorski R.: Theory of distributions. Elsevier, 1973

5. Baumgarten D., Braunss G., Wagner O.: An extension of the Gel'fand-Shilov regularization method and its application to the construction of causal solutions for nonlinear wave equations. Preprint, 1975

6. Bogoliubov N.N., Parasiuk O.S.: Ueber die Multiplikation der Kausalfunktionen in der Quantentheorie der Felder. Acta Math., 97,1957, 227-266

7. Bollini C.G., Giambiagi J.J., Gonzales Dominguez A.: Analytic regularization and the divergencies of quantum field theories. Nuovo Cim., 31,3,1964, 550-561

8. Braunss G.: On the regularization of functional equations. Math. Ann., 186,1970 70-80

9. Braunss G.: Causal functions of nonlinear wave equations. J. Diff. Eq., 9,1, 1971, 86-92

10. Braunss G.: Weak solutions and junction conditions of a certain class of nonlinear partial differential equations. Preprint 1971

11. Braunss G., Liese R.: Canonical products of distributions and causal solutions of nonlinear wave equations. J. Diff. Eq., 16,1974, 399-412

12. Bredimas A.: Sur une theorie electromagnetique non lineaire. Lett. Nuovo Cim., 6,14,1973, 569-572

13. Bredimas A.: Generalized convolution product and canonical product between distributions. Some simple applications to physics. Lett. Nuovo Cim., 13,16, 1975, 601-604

14. Bredimas A.: La differentiation d'ordre complexe, le produit de convolution generalise et le produit canonique pour les distributions. C.R. Acad. Sc. Paris, 282,1976, A, 37-40

15. Bredimas A.: Applications en Physique du theoreme d'inversion de la transformation spherique de Radon dans R^n. C.R. Acad. Sc. Paris, 282,1976, A, 1175-1178

16. Bredimas A.: Extensions, proprietes complementaires et applications des operateurs de differentiation a gauche et a droite d'ordre complexe. C.R. Acad. Sc. Paris, 283,1976, A, 3-6

17. Bredimas A.: Applications a certaines equations differentielles des premier et second ordre a coefficients polynomiaux des operateurs de differentiation d'ordre complexe a gauche et a droite. C.R. Acad. Sc. Paris, 283,1976, A, 337-340

18. Bredimas A.: La differentiation d'ordre complexe et les produits canoniques et la convolution generalise. C.R. Acad. Sc. Paris, 283,1976, A, 1095-1098

19. Bremermann H.J., Durand L.: On analytic continuation, multiplication and Fourier transform of Schwartz distributions. J. Math. Phys., 2,2,1961, 240-258

20. Bremermann H.J.: On finite renormalization constants and the multiplication of causal functions in perturbation theory. Preprint, 1962

21. Bremermann H.J.: Some remarks on analytic representations and products of distributions. SIAM J. Appl. Math., 15,4,1967, 929-943

22. Burrill C.W.: Foundations of Real Numbers. Mc Graw Hill, 1967

23. Carmignani R., Schrader K.: Subfunctions and distributional inequalities. SIAM J. Math. Anal., 8,1,1977, 52-68

24. Conley C.C., Smoller J.A.: Shock waves as limits of progressive wave solutions of higher order equations. Comm. Pure Appl. Math., 24,1971, 459-472

25. Conway E.D.: The formation and decay of shocks for a conservation law in several dimensions. Arch. Rat. Mech. Anal., 64,1,1977, 47-57

26. Dafermos C.M.: Quasilinear hyperbolic systems that result from conservation laws. Leibovich S., Seebass A.R. (ed.): Nonlinear Waves, Cornell Univ. Press, 1974, 82-102

27. Davies E.B.: Eigenfunction expansions for singular Schroedinger operators. Arch. Rat. Mech. Anal., 63,3,1977, 261-272

28. Deift P., Simon B.: On the decoupling of finite singularities from the question of asymptotic completeness in two body quantum systems. Preprint, 1975

29. De Jager E.M.: Divergent convolution integrals in electrodynamics. ONR, Tech. Rep. Dept. Math. Univ. Calif. Berkeley, 1963

30. De Mottoni P., Texi A.: On distribution solutions for nonlinear differential equations: nontriviality conditions. J. Diff. Eq., 24,1977, 355-364

31. Di Perna R.J.: Singularities of solutions of nonlinear hyperbolic systems of conservation laws. Arch. Rat. Mech. Anal., 60,1976, 75-100

32. Douglis A.: An ordering principle and generalized solutions of certain quasi linear partial differential equations. Comm. Pure Appl. Math., 12,1959, 87-112

33. Edwards R.E.: Functional Analysis. Holt, 1965

34. Ehrenpreis L.: Solutions of some problems of division. Amer. J. Math., 76, 1954, 883-903

35. Fisher B.: The product of distributions. Quart. J. Math. Oxford, 22,2,1971, 291-298

36. Fisher B.: The product of the distributions $x_+^{-r-1/2}$ and $x_-^{-r-1/2}$. Proc. Camb. Phil. Soc. 71,1972, 123-130

37. Fisher B.: The product of the distributions x^{-r} and $\delta^{(r-1)}(x)$. Proc. Camb. Phil. Soc. 72,1972, 201-204

38. Fisher B.: Some results on the product of distributions. Proc. Camb. Phil. Soc 73,1973, 317-325

39. Fisher B.: Singular products of distributions. Math. Ann., 203,1973, 103-116

40. Fisher B.: The neutrix distribution product $x_+^{-r}\delta^{(r-1)}(x)$. Stud. Sci. Math. Hung., 9,1974, 439-441

41. Fisher B.: Distributions and the change of variable. Bull. Math. Soc. Sci. Math. Rom., 19,1975, 11-20

42. Fleishman B.A., Mahar T.J.: Boundary value problems for a nonlinear differential equation with discontinuous nonlinearities. Math. Balkanica, 3,1973, 98-108

43. Fleishman B.A., Mahar T.J.: A perturbation method for free boundary problems of elliptic type. Preprint, 1976

44. Flugge S.: Practical quantum mechanics I. Die Grundlehren der Mathematischen Wissenschaften. Band 177, Springer, 1971

45. Fountain L., Jackson L.: A generalized solution of the boundary value problem. Pacif. J. Math., 12,1962, 1251-1272

46. Fuchssteiner B.: Eine assoziative Algebra ueber einem Unterraum der Distributionen. Math. Ann., 178,1968, 302-314

47. Galambos J.: Representation of real numbers by infinite series. Springer Lecture Notes in Mathematics, vol. 502, 1976

48. Garabedian P.R.: An unsolvable equation. Proc. AMS, 25,1970, 207-208

49. Garnir H.G.: Solovay's axiom and functional analysis. Springer Lecture Notes in Mathematics, vol. 399, 1974, 189-204

50. Gel'fand J.M.: Some problems in the theory of quasilinear equations. Uspehi Mat. Nauk, 14,1959, 87-158

51. Glimm J., Lax P.D.: Decay of solutions of systems of nonlinear hyperbolic conservation laws. Memoirs of AMS, nr. 101, 1970

52. Golubitsky M., Schaeffer D.G.: Stability of shock waves for a single conservation law. Adv. Math., 15,1975, 65-71

53. Gonzales Dominguez A., Scarfiello R.: Nota sobre la formula v.p. $\frac{1}{x} \cdot \delta = -\frac{1}{2}\delta'$ Rev. Union Mat. Argen., 1, 53-67, 1956

54. Guettinger W.: Quantum field theory in the light of distribution analysis. Phys. Rev., 89,5,1953, 1004-1019

55. Guettinger W.: Products of improper operators and the renormalization problem of quantum field theory. Progr. Theor. Phys., 13,6,1955, 612-626

56. Guettinger W.: Generalized functions and dispersion relations in physics. Fortschr. Phys., 14,1966, 483-602

57. Guettinger W.: Generalized functions in elementary particle physics and passive system theory: recent trends and problems. SIAM J. Appl. Math., 15,4, 1967, 964-1000

58. Guckenheimer J.: Solving a single conservation law. Preprint, 1973

59. Harten A., Hyman J.M., Lax P.D., Keyfitz B.: On finite difference approximations and entropy conditions for shocks. Comm. Pure Appl. Math., 29,1976, 297-322

60. Hepp K.: Proof of the Bogoliubov-Parasiuk theorem on renormalization. Comm. Math. Phys. 2,1966, 301-326

61. Hirata Y., Ogata H.: On the exchange formula for distributions. J. Sc. Hirosh. Univ., 22,1958, 147-152

62. Hopf E.: The partial differential equation $u_t + uu_x = \mu u_{xx}$. Comm. Pure Appl. Math., 3,1950, 201-230

63. Horvat J.: An introduction to distributions. Amer. Math. Month., March 1970, 227-240

64. Hoermander L.: Linear Partial Differential Operators. Springer, 1963

65. Hoermander L.: An introduction to complex analysis in several variables. Van Nostrand, 1966

66. Itano M.: On the multiplicative product of distributions. J. Sc. Hirosh. Univ. 29,1965, 51-74

67. Itano M.: On the multiplicative products of x_+^α and x_+^β. J. Sc. Hirosh. Univ. 29,1965, 225-241

68. Itano M.: On the theory of the multiplicative products of distributions. J. Sc. Hirosh. Univ., 30,1966, 151-181

69. Itano M.: Remarks on the multiplicative product of distributions. Hirosh. Math. j., 6,1976, 365-375

70. Jeffrey A.: The propagation of weak discontinuities in quasilinear symmetric hyperbolic systems. J. Appl. Math. Phys., 14,1963, 301-314

71. Jeffrey A.: Quasilinear hyperbolic systems and waves. Pitman, 1976

72. Jelinek J.: Sur le produit simple de deux distributions. Comm. Math. Univ. Carol., 5,4,1964, 209-214

73. Kang H., Richards J.: A general definition of convolution for distributions. Preprint, 1975

74. Kang H.: A general definition of convolution for several distributions. Preprint, 1975

75. Keisler H.J.: Elementary Calculus. Prindle, 1976

76. Kelemen P.J., Robinson A.: The nonstandard $\lambda : \Phi_2^4(x)$: model. I. The technique of nonstandard analysis in theoretical physics. II. The standard model from a nonstandard point of view. J. Math. Phys., 13,12,1972, 1870-1878

77. Kelemen P.J.: Quantum mechanics, quantum field theory and hyper quantum mechanics. Springer Lecture Notes in Mathematics, vol. 369, 1974, 116-121

78. Keller K.: Irregular operations in quantum field theory I, II, III, IV, V Preprint, 1977

79. Keyfitz Quinn B.: Solutions with shocks: an example of an L_1 contractive semigroup. Comm. Pure Appl. Math., 24,1971, 125-132

80. Knops R.J.: Comments on nonlinear elasticity and stability. Springer Lecture Notes in Mathematics, vol. 564, 1976, 271-290

81. Koenig H.: Multiplikation von Distributionen. Math. Ann., 128,1955, 420-452

82. Koenig H.: Multiplikation und Variablentransformation in der Theorie der Distributionen. Arch. Math., 6,1955, 391-396

83. Koenig H.: Multiplikationstheorie der verallgemeinerten Distributionen. Bayer. Akad. Wiss. Math. Nat. Kl. Abh. (N.F.), 82,1957

84. Larra Carrero L.: Stability of shock waves. Springer Lecture Notes in Mathematics, vol. 564, 1976, 316-328

85. Laugwitz D.: Eine Einfuehrung der δ Funktionen. Bayer. Akad. Wiss. Math. Naturw. Kl. 1959, 41-59

86. Laugwitz D.: Anwendungen unendlich kleiner Zahlen. I. Zur Theorie der Distributionen. II. Ein Zugang zur Operatorenrechnung von Mikusinski. J.f.d. reine und angew. Math., 207,1961, 53-60, 208,1961, 22-34

87. Laugwitz D.: Bemerkungen zu Bolzano Groessenlehren. Arch. Hist. Ex. Sc., 2, 1965, 398-409

88. Laugwitz D.: Eine nichtarchimedische Erweiterung angeordneter Koerper. Math. Nacht., 37,1968, 225-236

89. Lax P.D.: Weak solutions of nonlinear hyperbolic equations and their numerical computation. Comm. Pure Appl. Math., 7,1954, 159-193

90. Lax P.D.: Hyperbolic systems of conservation laws and the mathematical theory of shock waves. SIAM Regional Conference Series in Appl. Math., nr. 11, 1973

91. Leibovich S., Seebass A.R. (ed.): Nonlinear Waves. Cornell Univ. Press, 1974

92. Levine H.A., Payne L.E.: Nonexistence of global weak solutions for classes of nonlinear wave and parabolic equations. J. Math. Anal. Appl., 55,1976, 329-334

93. Lewy H.: An example of a smooth linear partial differential equation without solution. Ann. Math., 66,2,1957, 155-158

94. Ligeza J.: On generalized solutions of some differential nonlinear equations of order n . Ann. Pol. Math., 31,1975, 115-120

95. Lightstone A.H.: Infinitesimals. Amer. Math. Month., March 1972, 242-251

96. Lightstone A.H., Kam Wong: Dirac delta functions via nonstandard analysis. Canad. Math. Bull., 81,5,1975, 759-762

97. Lightstone A.H., Robinson A.: Non Archimedean fields and asymptotic expansions. North Holland, 1975

98. Liverman T.P.G.: Physically motivated definition of distributions. SIAM J. Appl. Math., 15,1967, 1048-1076

99. Lojasiewicz S.: Sur la valeur d'une distribution dans un point. Bull. Acad. Polon. Sc., 4,1956, 239-242

100. Lojasiewicz S.: Sur la valeur et la limite d'une distribution dans un point. Stud. Math., 16,1957, 1-36

101. Lojasiewicz S.: Sur la fixation des variables dans une distribution. Stud. Math., 17,1958, 1-64

102. Luxemburg W.A.J.: What is Nonstandard Analysis? Amer. Math. Month., 80,6, part II, 38-67

103. Malgrange B.: Existence et approximation des solutions des equations aux derivees partielles et des equations de convolution. Ann. Inst. Fourier Grenoble, 6,1955-56, p.271-375

104. Meisters G.H.: Translation invariant linear forms and a formula for the Dirac measure. J. Funct. Anal., 8,1971, 173-188

105. Mikusinski J.: Sur la methode de generalisation de Laurent Schwartz et sur la convergence faible. Fund. Math., 35,1948, 235-239

106. Mikusinski J.: Irregular operations on distributions. Stud. Math., 20,1961, 163-169

107. Mikusinski J.: Criteria of the existence and of the associativity of the product of distributions. Stud. Math., 21,1962, 253-259

108. Mikusinski J.: On the square of the Dirac delta distribution. Bull. Acad. Pol. Sc., 14,9,1966, 511-513

109. Mikusinski J.: Germs and their operational calculus. Stud. Math., 26,1966, 315-325

110. Mikusinski J., Sikorski R.: The elementary theory of distributions I, II. Rozprawy Matematyczne, 12,1957 and 25,1961

111. Moss W.F.: A remark on convergence of test functions. J. Austral. Math. Soc. 20, A, 1975, 73-76

112. Nokanishi N.: Complex dimensional invariant delta functions and lightcone singularities. Comm. Math. Phys., 48,1976, 97-118

113. Oleinik O.A.: On the uniqueness of the generalized solution of the Cauchy problem for a nonlinear system of equations occuring in mechanics. Uspehi Mat. Nauk., 78,1957, 169-176

114. Palamodov V.P.: Linear Differential Operators with Constant Coefficients. Springer, 1970

115. Pearson D.B.: An example in potential scattering illustrating the breakdown in asymptotic completeness. Comm. Math. Phys., 40,1975, 125-146

116. Pearson D.B.: General theory of potential scattering with absorption at local singularities. Preprint, 1975

117. Plesset M.S., Prosperetti A.: Bubble dynamics and cavitation. Ann. Rev. Fluid Mech., 9,1977, 145-185

118. Pogorelov A.V.: Monge Ampere equation of elliptic type. Noordhoff, Groningen, 1964

119. Pogorelov A.V.: On a regular solution of the n dimensional Minkowski problem. Soviet Math. Dokl., 12,1971, 1192-1196

120. Pogorelov A.V.: The Dirichlet problem for the n dimensional analogue of the Monge Ampere equation. Soviet Math. Dokl., 12,1971, 1727-1731

121. Portnoy S.L.: On solutions to $u_t = \Delta u + u^2$ in two dimensions. J. Math. Anal. Appl., 55,1976, 291-294

122. Reed M.C.: Abstract Nonlinear Wave Equations. Springer Lecture Notes in Mathematics, vol. 507, 1976

123. Richtmyer R.D.: On the structure of some distributions discovered by Meisters. J. Funct. Anal., 9,1972, 336-348

124. Robinson A.: Nonstandard Analysis. North Holland, 1966

125. Rosinger E.E.: Embedding the $D'(R^n)$ distributions into pseudotopological algebras. Stud. Cerc. Mat., 18,5,1966, 687-729

126. Rosinger E.E.: Pseudotopological spaces. Embedding the $D'(R^n)$ distributions into algebras. Stud. Cerc. Mat., 20,4,1968, 553-582

127. Rosinger E.E.: A distribution multiplication theory. Preprint, 1974

128. Rosinger E.E.: Division of distributions. Pacif. J. Math., 66,1,1976, 257-263

129. Rosinger E.E.: An associative, commutative distribution multiplication I, II. Preprint, 1976

130. Rosinger E.E.: Nonsymmetric Dirac distributions in scattering theory. Springer Lecture Notes in Mathematics, vol. 564, 1976, 391-399

131. Rosinger E.E.: Nonlinear shock waves and distribution multiplication. Preprint, 1977

132. Sabharwal C.L.: Multiplication of singularity functions by discontinuous functions in Schwartz distribution theory. SIAM J. Appl. Math., 18,2,1970, 503-510

133. Schaeffer D.G.: A regularity theorem for conservation laws. Adv. Math., 11,3, 1973, 368-386

134. Schmieden C., Laugwitz D.: Eine Erweiterung der Infinitesimalrechnung. Math. Zeitschr., 69,1958, 1-39

135. Schwartz L.: Sur l'impossibilite de la multiplication des distributions. C.R. Acad. Sc. Paris, 239,1954, 847-848

136. Shiraishi R., Itano M.: On the multiplicative products of distributions. J. Sc. Hirosh. Univ., 28,1964, 223-235

137. Shiraishi R.: On the value of distributions at a point and the multiplicative products. J. Sc. Hirosh. Univ., 31,1967, 89-104

138. Shiu Yuen Cheng, Shing Tung Yau: On the regularity of the Monge Ampere equation. Comm. Pure Appl. Math., 30,1977, 41-68

139. Sikorsky R.: A definition of the notion of distribution. Bull. Acad. Pol. Sc. 2,5,1954, 209-211

140. Simon B.: Quantum mechanics for Hamiltonian defined as quadratic forms. Princeton Univ. Press, 1971

141. Slowikowski W.: A generalization of the theory of distributions. Bull. Acad. Pol. Sc., 3,1,1955, 3-6

142. Slowikowski W.: On the theory of operator systems. Bull. Acad. Pol. Sc., 3,3, 1955, 137-142

143. Speer E.R.: Analytic renormalization. J. Math. Phys., 9,9,1968, 1404-1410

144. Stroyan K.D., Luxemburg W.A.J.: Introduction to the theory of infinitesimals. Acad. Press, 1976

145. Struble R.A.: An algebraic view of distributions and operators. Stud. Math., 37,1971, 103-109

146. Stuart C.A.: Differential equations with discontinuous nonlinearities. Arch. Rat. Mech. Anal., 63,1,1976, 59-75

147. Stuart C.A.: Boundary value problems with discontinuous nonlinearities. Springer Lecture Notes in Mathematics, vol. 564, 1976, 472-484

148. Szaz A.: Convolution multipliers and distributions. Pacif. J. Math., 60,2,1975 267-275

149. Tavera G., Burnage H.: Sur les ondes de choc dans les ecoulements instationnaires. C.R. Acad. Sc. Paris, 284,1977, A, 571-573

150. Temple G.: Theories and applications of generalized functions. J. London Math. Soc., 28,1953, 134-148

151. Thurber J.K., Katz J.: Applications of fractional powers of delta functions. Springer Lecture Notes in Mathematics, vol. 369, 1974, 272-302

152. Tillmann H.G.: Darstellung der Schwartz'schen Distributionen durch analytische Funktionen. Math. Z., 77,1961, 106-124

153. Treves F.: Locally Convex Spaces and Linear Partial Differential Equations. Springer, 1967

154. Treves F.: Applications of distributions to PDE theory. Amer. Math. Month., March 1970, 241-248

155. Treves F.: Linear Partial Differential Equations. Gordon, Breach, 1970

156. Treves F.: Basic Linear Partial Differential Equations. Acad. Press, 1975

157. Van Osdol D.H.: Truth with respect to an ultrafilter or how to make intuition rigorous. Amer. Math. Month., April 1972, 355-363

158. Van Rootselaar B.: Bolzano's theory of real numbers. Arch. Hist. Exact Sc., 2,1964, 168-180

159. Vol'pert A.J.: On differentiation and quasilinear operators in the space of functions whose generalized derivatives are measures. Soviet Math. Dokl. 7,6,1966, 1586-1589

160. Vol'pert A.J.: The spaces BV and quasilinear equations. Math. USSR Sbornik, vol. 2,2,1967, 225-267

161. Zel'dovich Ja. B.: Hydrodynamics of the universe. Ann. Rev. Fluid Mech., 9, 1977, 215-228

162. Walter H.F.: Ueber die Multiplikation von Distributionen in einem Folgenmodell Math. Ann., 189,1970, 211-221

163. Whitham G.B.: Linear and Nonlinear Waves. Wiley, 1974